VOL. 05
PRACTICE

Edited by
Gavin Van Horn, Robin Wall Kimmerer, John Hausdoerffer

Center for Humans and Nature Press

Center for Humans and Nature Press, Libertyville 60030
© 2021 by Center for Humans and Nature

All rights reserved. No part of this book may be used or reproduced in any manner whatsoever without written permission, except in the case of brief quotations in critical articles and reviews.

For more information, contact the Center for Humans and Nature Press, 17660 West Casey Road, Libertyville, Illinois 60048.
Printed in the United States of America.

Cover and slipcase design: LimeRed, https://limered.io

ISBN-13: 978-1-7368625-0-6 (paper)
ISBN-13: 978-1-7368625-1-3 (paper)
ISBN-13: 978-1-7368625-2-0 (paper)
ISBN-13: 978-1-7368625-3-7 (paper)
ISBN-13: 978-1-7368625-4-4 (paper)
ISBN-13: 978-1-7368625-5-1 (set/paper)

Names: Van Horn, Gavin, editor | Kimmerer, Robin Wall, editor | Hausdoerffer, John, editor
Title: Kinship: belonging in a world of relations, vol. 5, practice / edited by Gavin Van Horn, Robin Wall Kimmerer, and John Hausdoerffer

Description: First edition. | Libertyville, IL: Center for Humans and Nature Press, 2021 | Identifiers: LCNN 2021909501 | ISBN 9781736862544 (paper)

Copyright and permission acknowledgments appear on page 148.

Center for Humans and Nature Press
17660 West Casey Road, Libertyville, Illinois 60048

www.humansandnature.org

Printed by Graphic Arts Studio, Inc. on Rolland Opaque paper. This paper contains 30% post-consumer fiber, is manufactured using renewable energy - Biogas and is elemental chlorine free. It is Forest Stewardship Council® and Rainforest Alliance certified.

CONTENTS

Gavin Van Horn
Kinning: Introducing the Kinship Series | 1

Heather Swan
After | 12

Thomas Lowe Fleischner
Natural History as a Practice of Kinship | 13

Amba Sepie
Settled Kin: Coming Home to
Where We Now Belong | 22

Kyle Powers Whyte
An Ethic of Kinship | 30

Matthew Hall
Kinship with Plants | 39

Nickole Brown
Mercy | 48

Ajay Rastogi
Entangled Kin: Merging Biotic
and Abiotic Beings | 53

Trebbe Johnson
The Coal Remembers | 63

Maya Ward
Getting in on the Making | 73

Tiokasin Ghosthorse
Nurturing Thoughts | 82

Alison Hawthorne Deming
Kinship with Trees and Crows | 90

Sunil Chauhan
Forest: A Festival of Friends | 94

Anthony Zaragoza and María Isabel Morales
A Pedagogy of Human Dignity and
Classroom Kinship | 104

Alison Hawthorne Deming
The Vocation of Care | 115

Jill Riddell
The Invitation | 118

John Hausdoerffer, Robin Wall Kimmerer,
Sharon Blackie, Enrique Salmón, Orrin Williams,
María Isabel Morales
Epilogue—Attention, Curiosity, Play,
Gratitude: Practices of Kinship | 127

Permissions | 148

Acknowledgments | 149

Contributors | 152

KINNING: INTRODUCING THE KINSHIP SERIES

Gavin Van Horn

The lines hung lightly suspended in midair. Twinkling, illuminated from within. Then vanished. I inclined my head. The lines reappeared, seeming to materialize out of emptiness. These weren't merely lines; they were radial strands precisely strung from a central axis, intricately woven. The glow from a nearby streetlamp caught within them, revealing a sacred geometry. I leaned closer. With a slight tilt of my face—altering the angle between my eyes and the spider's evening project—the lines alternately disappeared or disclosed themselves. Their creator, deftly putting the finishing touches on this work of body art, was smaller than my thumbnail. Her handiwork sparkled with its own radiance. For a moment I felt envious—then grateful. The craft on display demonstrated skills of which I was completely and utterly incapable. I drew in closer to get a better look at the stitching, which would likely hang for the evening before being pulled apart by a strong morning wind.

Kinship: Belonging in a World of Relations can be described as a series of books, five volumes that group different essays and poems according to the scale of their subject matter, from the composition of the cosmos to the gestures of the everyday: Planet, Place, Partners, Persons, Practice. But these books could just as easily be described as a web, a meshwork whose strands gather, crisscross, and link together a vast variety of subjects and experiences. Each book reaches beyond its pages, spinning silk filaments through the others; turn your head at the right angle and an intricate web appears—functional, sensorial, and artful.

The essays and poems you hold in your hands comprise lines, strands of ink, patterns on paper. In your imagination, these words may come to life, recalling and revealing shared relations with our fellow Earthlings—our kinfolk—who come in all shapes and sizes, from the bacterium swimming in your belly or lying on the tip of your tongue to the vibrant collective breath that sweeps across your face and into your lungs. Worth thinking about—and perhaps *thanking* about—are the shared threads between kinfolk, especially plantfolk, that make this breath exchange possible. Your life, my life, all of our lives depend on the quality of relations between us—the air we breathe, the water we drink, the food we eat and the food we become—within an exuberant, life-generating planetary tangle capable of nurturing intelligences that can spin webs and words.

Kinning

The words in this *Kinship* web gesture beautifully toward the relations—vital, wild processes—that are always present yet not always visible. Because these relations may be difficult to apprehend, it may seem as though the world is merely a collection of inert objects, full of nouns. You are you. I am me. That bird at the feeder is, sadly, referred to as "it." That river underneath the bridge and that mountain on the horizon are designated "natural resources." Some of us have rights, legal standing, personhood. Some of us—depending on which nation-state we happen to dwell in—don't.

Nouns have utility, yet they can mislead, perhaps even reifying the idea that the world is composed of things—some small, some large, some shiny, some dull, some with wings, some with legs, some with leaves, some with fur. This language-induced reduction would suggest everything is mere matter, a gathering of atoms, in more or less complicated geometries. Note, however, even in that last sentence, an interesting gerund slipped into the mix. What are atoms but a gathering of *relations*? What is a gathering of relations if not *relatives relating*? Just as when I tilted my head and the spiderweb "appeared," due to a slight change of perspective, it is possible with a shift of perspective to see the threads connecting worlds, all the relations that make us kin. The point here is one that comes up repeatedly in *Kinship*: Earth—and everything within it, including all that creates what we call earth—is a verb. All is in motion; all is relating.

The English language is noun dominant, and in comparison to many Indigenous languages, the animacy and agency of other beings and processes often receive less emphasis. Though obviously still relying on English, these *Kinship* volumes—because of their subject matter—challenge this object obsession at its core. *Kinship* can be considered a noun, of course, a state of being—whether this is couched in terms of biological genetics; family, clan, or species affiliation; shared and storied relations and memories that inhere in people and places; or more metaphorical imaginings that unite us to faith traditions, cultures, countries, or the planet. But the voices in these volumes point us toward an alternative perspective: kinship *as a verb*.

Perhaps this kinship-in-action should be called kinning. Humans are born kin, in any number of ways. But the words in this *Kinship* anthology collectively express something more than birthright claims: they point toward how it is possible to *become* kin. In this understanding, being kin is not so much a given as it is an intentional process. Kinning does not depend upon genetic codes. Rather, it is cultivated by humans, as one expression of life among many, many, many others, and it revolves around an ethical question: how to rightly relate? We are kinning as we (re)connect our bodies, minds, and spirits within a world that is not merely a collection of objects but "a communion of subjects," as Thomas Berry put it.[1] The essays and poetry in these volumes, at different scales and in different geographies, show possibilities for becoming better kin—more receptive to the languages of others, especially nonhuman others, and better listeners to their stories, which reach out to us through place and time. This vibrant world, as well as these volumes, offers invitations for kinning—practices of belonging with and amid our fellow earthly kin.

Three Threads in the Web

Three conceptual threads came together to inform and inspire the creation of the multiple *Kinship* volumes. I'd like to briefly mention them here, as readers may want to be alert to them throughout various essays in the books.[2]

The first thread is a cosmovision—and, increasingly, legal recognition—that acknowledges and understands nonhumans, including entire watersheds, forests, or mountains, as persons. In the West, many of us have inherited a settler-colonial worldview that uplifts the human individual—often implicitly or explicitly identified with whiteness and maleness—as the locus of meaning and center of importance while reducing nature to resources, property, or fungible commodities. Bootstrap economics and the literary hero's journey reinforce such thinking—the lone figure encountering and overcoming obstacles, conquering beasts, and emerging victorious above the fracas. From this vantage, there are human persons (and now corporate "persons"), and there is everything else.

The religious studies scholar Graham Harvey, in his wide-ranging study of animistic and neo-animistic cultures and movements, upends such notions. Harvey observes that from an animistic perspective, "the world is full of persons, only some of whom are human." Persons, he goes on to write, are not equated solely with human beings in many cultures. Rather, the term serves as a broader umbrella for those beings who are perceived as displaying agency (and this encompasses landscapes, rocks, and bodies of water, in addition to plants and nonhuman animals):

> Persons are beings, rather than objects, who are animated and social towards others (even if they are not always sociable).

Animism *may* involve learning how to recognize who is a person and what is not—because it is not always obvious and not all animists agree that everything that exists is alive or personal. However, animism is more accurately understood as being concerned with learning how to be a good person in respectful relationships with other persons.[3]

When I first read Harvey fifteen years ago or so, I didn't know how or if this type of cosmology could ever make its way into mainstream Western consciousness. Then, in March 2017, the Whanganui River (Te Awa Tupua), the third-largest river in Aotearoa New Zealand, grabbed international headlines. The Whanganui officially gained legal status as a living entity with the same rights of personhood as a human being. More than a change in legal nomenclature, this reclassification of the river stands as a significant bicultural effort to bring disparate systems of law and care together among New Zealanders of European descent and the native Māori population (Te Āti Haunui-a-Pāpārangi).[4]

The designation of the Whanganui River is one instance in a growing number of cases in which legal personhood is being granted to nonhuman entities. An overlapping set of localized and national governmental precedents, many of which involve personhood language, for example, began to gain traction in 2006 by focusing on the "rights of nature."[5] Ecuador and Bolivia both included rights-of-nature clauses in their national constitutions in 2008 and 2010, respectively. In Colombia, courts ruled in favor of personhood for the Amazon and Atrato Rivers. In 2016, the Ho-Chunk Nation in Wisconsin amended their tribal constitution to include rights-of-nature language: "Ecosystems, natural communities, and species within the Ho-Chunk Nation territory possess inherent, fundamental, and inalienable rights to naturally exist, flourish, regenerate, and evolve." In 2017, the Ponca Nation in Oklahoma recognized rights of nature as statutory law to combat fracking. Australia, India, and Nepal have also taken steps toward establishing rights of nature. In 2019, the Yurok Tribal Council passed a resolution that declared the personhood of the Klamath River in the Pacific Northwest. Such landmark legal and legislative actions represent efforts to give "voice" to other-than-human beings, ensuring their inherent rights to exist and flourish. Gerard Albert, the lead Maori negotiator on behalf of the Whanganui *iwi* (tribe), summed up this sense of responsibility well: "We can trace our genealogy to the origins of the universe. And therefore, rather than us being masters of the natural world, we are part of it. We want to live like that as our starting point. And that is not an anti-development, or anti-economic use of the river but to begin with the view that it is a living being, and then consider its future from that central belief."[6]

Recognition of kinship has many overlaps with these attempts to recognize personhood. The commonality lies in a respect for the agency of other beings and concerted efforts to treat them with dignity and even deference. This brings us to the second thread. *Kincentric ecology*, a phrase coined by ethnobotanist Enrique Salmón, provides a helpful guide for understanding kinship: an intertwining of the social, mythological, and practical. Salmón asserts that "life in any environment is viable only when humans view their surroundings as kin; that their mutual roles are essential for their survival."[7] This perspective stands in marked contrast to the familiar, if not predominant, human chauvinism toward other species in so many national sociopolitical systems. From a kinship perspective, the landscapes of which humans are a part—including rocks, rivers, oceans, prominent geographic features, and other nonhuman plant and animal persons—provide a shared sense of place and require appropriate human care and respect.

This kinship is deep and wide—and dwells within the human body. In the past century and a half, evolutionary and ecological sciences have brought additional insights to bear on what it means to be human. In only the past few decades, evolutionary models are being transformed by research into symbiotic mergers at the cellular level, horizontal gene transfer, and seemingly chimeric creatures that rely on cooperative relationships between species from entirely different "kingdoms" of life. Kinship, it would seem, is key to understanding biotic and abiotic entanglement. A kincentric ecology emerges from cultures that recognize the importance of humans in maintaining right relations in particular landscapes. Far from presupposing that humans are a degrading force, sullying whatever we might touch, a kincentric ecology expresses the view that humans can actually play keystone roles in our landscapes, creating mutual flourishing. In other words, human beings are not merely kin by biological relation, but it is entirely possible that human communities and cultures can be good kin, salutary ecological collaborators alongside and with our nonhuman family members.

The third thread that inspired this *Kinship* series, I'm happy to say, comes from coeditor Robin Wall Kimmerer. Robin's work draws from scientific training and Indigenous knowledge in complementary ways. In *Braiding Sweetgrass: Indigenous Wisdom, Scientific Knowledge and the Teachings of Plants*, she explores her own history of loss and recovery as a member of the Citizen Potawatomi Nation and describes how Indigenous perspectives can transform engagement with a living world. Perhaps nowhere is this clearer than when she contrasts the "grammar of animacy" embedded in the Potawatomi language to conventional English and its objectifying pronouns. She makes a convincing case that an ethical revolution might depend on a language revolution. Finding ways to properly and respectfully acknowledge *ki* (the pronoun Robin proposes for our other-than-human kin) is a good place to begin.

You will recognize these three threads—nonhuman personhood, humans as relational participants in local ecologies, and the care expressed when addressing and engaging with our kinfolk through language—winding their way through all the volumes of *Kinship*.

Five Scales of Kinning

With all the amazing contributors gathered in this *Kinship* web, it might help the reader to know what we as editors were asking of them. The following are the questions we posed to our contributors for each *Kinship* volume, which feature the ways kinship can be understood at different scales: from deep time cosmic and evolutionary relationships; to community watersheds, landscapes, and bioregions; to interspecies engagements and mythological perceptions; to biological and symbolic understandings of human interbeing; to which kinds of practices are appropriate for making and becoming kin:

> *Volume 1: Planet*—With every breath, every sip of water, every meal, we are reminded that our lives are inseparable from the life of the world—and the cosmos—in ways both material and spiritual. What are the sources of our deepest evolutionary and planetary connections, and of our profound longing for kinship?
>
> *Volume 2: Place*—Given the place-based circumstances of human evolution and culture, global consciousness may be too broad a scale of care for us. To what extent does crafting a deeper connection with Earth's bioregions reinvigorate a sense of kinship with the place-based beings, systems, and communities that mutually shape one another?
>
> *Volume 3: Partners*—How do cultural traditions, narratives, and mythologies shape the ways we relate, or not, to other beings as kin? How do relations between and among different species foster a sense of responsibility and belonging in us?
>
> *Volume 4: Persons*—Kinship spans the cosmos, but it is perhaps most life changing when experienced directly and personally. Which experiences expand our understanding of being human in relation to other-than-human beings? How can

we respectfully engage a world full of human and nonhuman persons?

Volume 5: Practice—From the perspective of kinship as a recognition of nonhuman personhood, of kincentric ethics, and of *kinship* as a verb involving active and ongoing participation, how are we to live? What are the practical, everyday, and lifelong ways we *become* kin?

We invited our contributors because of their experiences, their expertise, their diverse backgrounds and geographical locations, and because of the way they've made kin with particular species—or some combination of all of these. We also invited our contributors to share their words because of their abilities to tell a good story.

In many of the essays in *Kinship*, there are statistics, references to academic sources, explorations of complicated ideas, and endnotes that may take a reader onward, but we above all wanted readers to hear people's stories. As human beings, we are storytelling animals. We lean a little closer when we overhear someone else say, "Oh my, have I got a good story for you!" For similar reasons, our Paleolithic ancestors likely leaned into the firelight—as it crackled against the cave walls at Lascaux, France, or Sulawesi, Indonesia—watching the aurochs, bison, horses, and deer dance before their eyes, or the warty pig and the babirusa (pig-deer).[8]

As storytelling creatures, when thinking of our personal relationships with the natural world, we may be predisposed to be on the lookout for epiphanies—the holy overwhelm, the big payoff, the road-to-Damascus moment, the final boss battle. But becoming kin, as the various stories in these volumes attest, consists of repeated intimacies, familiar encounters, and daily undoings and transformations that are dependent on visitations and conversations within a smaller circle of place. Awe-filled moments of raw contact with forces that relativize human importance should not be disparaged or discounted. Such experiences may stir deep wells of gratitude. But knowing that humans are relatives, responsible for the lives of others, as others are responsible for ours, remains at the heart of these volumes. If humans are relatives relating, not merely in terms of an abstract genetic code but as intimate familiars, the question then becomes how we as individual persons and communities might better cultivate these relations. How can we uproot the desire to impose our will upon the living worlds around us? How do we become more receptive to nonhuman languages and ways of being?

One step in this direction is the recognition that nature is not a passive object, a text awaiting our interpretation or exegesis, a thing humans approach solely for insights, entertainment, and "resources." The world all of us are part of and participate in is a relational exchange—alive, wildly generative, an ongoing conversation of bodies, desires, conflicts, and collaborations. There is no pinnacle here for humans to sit atop and gaze upon the masses. *Kinship* culminates, in volume 5, with a wide-ranging conversation about practices and ethics that embrace a world of other-than-human persons as worthy of our active care, concern, and respect. At a time when human fidelity to the natural world seems to be fraying, *Kinship* offers stories of solidarity, highlighting the deep interdependence that exists between humans and the more-than-human world. It explores challenging questions, including how communities might fairly and effectively give voice to nonhuman beings and landscapes. And it highlights the cosmologies, mythical narratives, and everyday practices that embrace a world of other-than-human persons as worthy of response and responsibility.

Humans will survive and continue to tell our stories if we learn how to live well with our kin. The voices included in these volumes—these webs of words—and the collective wisdom they express, invite us into this kind of kinning. These are the stories of how to listen to voices other than our own.

Lean a little closer into the firelight. Up on the ceiling of the cave, or near the streetlamp, or between branches in the forest, or in the corner of the room within which you are currently reading, a web may be flashing in a flicker of light.

NOTES

1. Thomas Berry, *Evening Thoughts: Reflecting on Earth as Sacred Community* (San Francisco: Sierra Club, 2006), 149. More on Berry as "geologian" can be found at the website of the Thomas Berry Foundation, http://thomasberry.org/life-and-thought/about-thomas-berry/geologian.
2. In addition to the themes, the three persons mentioned all have essays in the *Kinship* volumes. Graham Harvey, "Academics Are Kin, Too: Transformative Conversations in the Animate World," appears in *Vol. 4: Persons*; Enrique Salmón, "A Heart Rooted in Place: Poetic Dentists and Getting Rained On," is in *Vol. 2: Place*; and Robin Wall Kimmerer, "A Family Reunion near the End of the World," is in *Vol. 1*: Planet.
3. Graham Harvey, A*nimism: Respecting the Living World* (New York: Columbia University Press, 2006), xi.
4. See Anna M. Gade, "Managing the Rights of Nature for Te Awa Tupua," *Edge Effects*, September 5, 2019, https://edgeeffects.net/te-awa-tupua/. As my coeditor Robin pointed out to me, "This can be understood, not [that] the river *gained* personhood" but "more that Western institutions came to acknowledge its intrinsic personhood, under the tutelage of the Maori, who have always recognized this inherent nature."
5. Craig M. Kauffman and Pamela L. Martin, "Constructing Rights of Nature Norms in the U.S., Ecuador, and New Zealand," *Global Environmental Politics* 18, no. 4 (2018): 43–62, https://www.mit-pressjournals.org/doi/pdf/10.1162/glep_a_00481.
6. E. A. Roy, "New Zealand River Granted Same Legal Rights as Human Being," *The Guardian*, March 16, 2017, https://www.theguardian.com/world/2017/mar/16/new-zealand-river-granted-same-legal-rights-as-human-being.
7. Enrique Salmón, "Kincentric Ecology: Indigenous Perceptions of the Human-Nature Relationship," *Ecological Applications* 10, no. 5 (2000): 1327–32, https://www.researchgate.net/profile/Enrique_Salmon/publication/242186767_Kincentric_Ecology_Indigenous_Perceptions_of_the_HumanNature_Relationship/links/5c34e542a6fdccd6b59c2aa1/Kincentric-Ecology-Indigenous-Perceptions-of-the-HumanNature-Relationship.pdf.
8. Leang Timpuseng is the name of the cave in Sulawesi, Indonesia, whose figurative art has been geochemically dated using recently developed techniques. The results pushed back the timeline on some of the earliest known figurative cave art, and examples of symbolic thinking, to more than 35,000 years ago. "Find early paintings, particularly figurative representations like animals, and you've found evidence for the modern human mind," writes Jo Marchant in "A Journey to the Oldest Cave Paintings in the World," *Smithsonian Magazine*, January–February 2016, https://www.smithsonianmag.com/history/journey-oldest-cave-paintings-world-180957685/. The timeline continues to lengthen; in 2021, from a karst system in Sulawesi, a date of 45,500 years ago was announced for the earliest known representational work of art. See Adam Brumm et al., "Oldest Cave Art Found in Sulawesi," *Science Advances* 7, no. 3 (January 13, 2021): https://advances.sciencemag.org/content/7/3/eabd4648.

AFTER

Heather Swan

    There among the silences
find the ghost tree—

    the split black branches making
fissures in the clearing.

    Watch as the fog dresses
and undresses the wounds,

    the suppuration of bark,
so raw underneath.

    The birds can find
no purchase.

    Scavenge the esker,
make a circle of stones,

    kneel down wreathed in
feather and bracken.

    Prepare to knit yourself
back into the world.

NATURAL HISTORY AS A PRACTICE OF KINSHIP

Thomas Lowe Fleischner

We clamber out of the rafts onto the dry terrace, heading up into the side canyon, where, it is rumored, ancient pictographs and giant cottonwoods await. The first week of June, almost noon, this treeless terrace radiates heat. Our group—a dozen naturalists of diverse backgrounds and ages, drawn together on this river voyage by a shared sense of adventure and inquisitiveness—pushes for the bend in the canyon, where we might finally gain some modicum of shade. While still in the harsh grip of the relentless sun, we're stopped in our tracks: we notice a living being atop the nondescript pile of rocks a hundred feet off the path. One after another, we gasp as we gaze through binoculars—startled by exquisite beauty. From a distance, this lizard appears little different from the rocks on which it sits. Seen close up, though, it is simply stunning: golden head and bright yellow feet; greenish back, spotted blue, ringed with golden stripes; brilliant orange patches; and chocolate brown-and-white patterning on a long tail, which droops off the edge of this jumbled sandstone platform. First, we gain clarity on species: Eastern Collared Lizard. Once we think through these details of color and pattern, we recognize this being as an adult male. "It" becomes "he."

He sits motionless, but our group erupts into ecstatic whoops, trying but failing to keep quiet. A few days later, a photo of this lizard on social media elicits hundreds more awed responses.

We are all so hungry for kinship, so ready to affiliate with the beauty that emerges when we bother to pay attention from the seemingly drab background of our lives.

Kinship. The sense of affiliation, of belonging. We all need it. But too commonly, we have lost this sense of connection in our human world: windows rolled up tight, locks pressed shut, children kept indoors, neighbors unmet. As for the multitude of worlds beyond the merely human, our lack of kinship is so thorough it often goes unnoticed. I once wrote that "our deepest affinity is for this rich and remarkable world we live in—our fellow beings, the textures and colors of landforms, the luscious scents of each place we touch."[1] This kind of expansive, interspecific affinity is deep in our bones, encoded in our genes.

But we live in an historical anomaly—human acknowledgment of the rest of the living world has never been so rare as it is today. Over the past few centuries, the dominant western culture of commerce has developed strategies to push this broader sense of kinship aside and foist upon us the tragic idea that connection with more-than-human nature is not worthy of adult attention. Yet capitalist impulses often dissolve in the presence of the innate, self-directed fascination—what's *this*?!—that we all were born with. Watch any small child anywhere, and you'll witness how deeply embedded our human curiosity about our world is—leaning down to turn over stones, stretching to peer into a bird's nest. Collective disregard of our inherent, full-on attentiveness to the world represents a momentous miscalculation, a massive plunge to the edge of a psychospiritual abyss.

It is critical that we break down barriers to affinity so that we can open up our sense of kinship. This is conspicuously true these days in human social dynamics, as we witness mass anxiety, despair at random violence, erect walls along borders, and cordon off neighborhoods with iron gates. But there is an even deeper need to transcend the eco-tribalism of our own species—the self-destructive notion that only we humans matter.

Our species must strive to reinhabit a world of broader and deeper connectivity and interpenetration. No task is more urgent, no effort more fundamentally human and humane: to enlarge our circle of affinity, our web of kinship. As the writer Scott Russell Sanders put it: "Our sense of moral obligation arises from a feeling of kinship. The illusion of separation . . . is the source of our worst behavior. The awareness of kinship is the source of our best behavior."[2]

One hundred fifty feet above the muddy floor of this tropical rain forest—snow-clad Andean peaks 150 miles in one direction, the Atlantic coast, where this surface water eventually flows, more than 3,000 miles in the other. Opal-Crowned Tanagers—smaller than my fist; luminescent cobalt plumage contrasting with a glowing stripe above the eye, and a patch of the same hue at the base of the tail—appear out of the receding rainfall of the canopy, descending like tiny feathered jewels into the welcoming watery cups of bromeliad flowers, filled to the brim by last night's downpour. One by one, these diminutive birds begin plashing themselves clean in the freshly captured rainwater, here in this habitat usually beyond the realm of humans. Yes, this moment represents data—a new species for a list. But any impulse of rationality is overpowered by something more primal—the sudden flush of awe, suffusing through my whole body: *these* gorgeous beings in *this* intimate encounter. The sense of this moment as a gift reverberates long after I have descended back to the forest floor and followed the faint trail back to where our canoe waits at the shore of the black-watered lake.

So many of our cultural assumptions work against connection and kinship. Indeed, our very language is structured to deny kinship with Others. The Native American ecologist Robin Wall Kimmerer has written: "In the absence of knowing the names of our neighbors the plants, we are compelled to refer to them with the ubiquitous pronoun 'it.' . . . 'It' robs a person of their humanity and reduces them to the lowly status of an object. And yet—in English, a being is either a human or a thing."[3] She goes on to assert that we need a new pronoun—one that denotes respect and animacy rather than objecthood. Drawing on her native Anishinaabe language, she suggests *ki* as a respectful pronoun for an animate being of the Earth. And the plural of *ki* already exists in English: *kin*. Thus, what might seem at first to be a linguistic contrivance turns out to lubricate the psychic gears of our turning toward kinship. As Kimmerer states, "The language of animacy, of kinship, can be medicine for a broken relationship."[4]

Words like *ki* can open up new possibilities. Words can also constrain experience. For example, the sterile, bureaucratic word *environment* is part of the problem for "environmentalists." Who can love such a dry term? The root of the word, *environ*, denotes surroundings, or simply what's around us. *Environment*, by its very nature, is vague—removed from, and less important than, us. It certainly does not prioritize a sense of kinship with the greater world. Simply referring to animate beings with respect, and acknowledging actual individual lives rather than abstract renderings of lives, goes a long way toward establishing a baseline of kinship in communication.

It is not hard to be distracted by *Penstemon* flowers. They come in several colors—scarlet, lavender, white tinged with pink—and all are tubular (the botanist would say "have fused corolla"), but the flower tube of some is dramatically elongated while in others it is scrunched-up and squat. Even more delight comes to those who look *inside* the flower. *Penstemon* is named for an anomaly in one of its stamens—the male part of the flower, a long filament capped by the pollen-bearing anther. In this genus, though, one of the five stamens differs from the other four: it lacks pollen but shows off other features instead. Different species exhibit distinct shapes and textures of this fifth stamen, the *staminode*—silky smooth in some, crowded with hairs in others. Just as the flowers display different colors on the exterior, so the inner forms present diversity, too. And a careful look inside this flower reveals another botanical truth: this being is neither he nor she, but both. When we pay attention, we find our social assumptions challenged, even more fully than they are in human political discourse. It turns out that in the plant world, plants of only one sex are very much in the minority. What is "normal" in nature can surprise us. In botany, *bisexual* and *perfect* are synonyms.

How do we rediscover passion for the world? What is required to build a sense of human community? Mutual respect, an opportunity for positive social interaction, and clear communication. The same ingredients—frequent interaction, honesty, and a strong sense of respect—undergird a healthy sense of belonging, of kinship, with the fuller community of life. What promotes frequent interaction with and respect for nonhuman Others? The practice of *natural history* creates a forum for interaction with Others, encouraging compassion and respect, helping us rediscover passion for the world, and for one another.

Natural history is the practice of falling in love with the world. Or, as I have defined it previously, "a practice of intentional, focused attentiveness and receptivity to the more-than-human world, guided by honesty and accuracy."[5] Natural history, then, is *practicing* attentiveness—a doing, a verb, not a noun.

The term *natural history—historia naturalis* in the original Latin—was coined by the Roman philosopher and writer Pliny the Elder in the same century that Jesus walked Earth. His *Historia naturalis*—literally, "the story of nature"—was the first encyclopedia, the first attempt to capture in writing everything known about the world around us: a multivolume compendium on plants, animals, minerals, stars, and a great deal more. From the beginning, then, natural history was expansive—broadly and deeply inquisitive. While the term *natural history* is two thousand years old, the practice of open-minded attentiveness goes back to the very origins of our species. Different contexts have provided different variants of natural history: curiosity cabinets in Victorian England, rows of shells in a seashore cabin, or a subset of scientific ecology in the world of twentieth-century research. But across the stretch of history, there has never been a moment in the story of human existence when natural history was so little practiced.[6]

Heat waves shimmer here at the desert's upper edge—the narrow ecotonal band where saguaro cacti and mesquite from below intermingle with junipers from the mesas above. Piquant seepwillow scent and the damp arroyo sand. Butterflies—blues, whites, admirals, and, especially, queens (think smaller, darker monarchs)—fountain up through willows along the length of this short canyon. The buoyance of many thousands of butterflies contrasts with the stark stillness of the hot, arid plain just beyond. This burst of life energy, oblivious to human concerns, transforms this arid landscape from a sere backdrop to a many-colored tapestry of delight.

And it helps me transcend the confines of my busily thinking mind into the rich realm of the unexpectable—often joyous, occasionally horrendous, always enlivening.

Attentive natural history helps us see and acknowledge more of the world. Watching birds at a backyard feeder, tracing the veins of rock with our fingertips, getting on hands and knees to look at the miracle of a spider's web, sitting back on a mountain peak and imagining the tectonic forces of creation and the glacial forces that sculpt the jagged ridges before us: there is literally no limit to what is presented before us each day, available for our attention. By its very nature, natural history practice extends our psyches beyond the limits of the purely human, into the realm of the greater psyche of the world. The field biologist Christopher Norment has described natural history field study as "sympathetic observation."[7] The research scientists Ron Pulliam and Nickolas Waser proclaim the importance of "natural history intuition."[8]

Along with a great many colleagues, my own work has been focused on promoting a renaissance of expansive, interdisciplinary natural history, fostering opportunities for people of all backgrounds to remember what it means to be in love with the world. For many years, this work took place from a professor's perch, leading students into the field, from Alaska to Mexico; Southwest canyons to Maine coast islands. More recently, I've been at the helm of a small nonprofit with a big mission—the Natural History Institute, which seeks to provide leadership and resources for a revitalized practice of natural history that integrates art, science, and the humanities to promote the health and well-being of humans and the rest of nature. This work involves public lectures, art exhibits, scientific research, and the convening of confluences of ideas. Sometimes, it takes new friends down a river to encounter the breathtaking surprise of a colorful lizard or elevates us into a rain-forest canopy to discover bathing tanagers.

A blurry backdrop becomes sharply etched, gains depth, becomes three-dimensional. And then, as one grasps the immense passages of time implied by the most conspicuous element of this landscape—the rocks—it becomes four-dimensional. Reddish rock and generic green transform into desert-varnished Navajo sandstone, fronted by Fremont Cottonwood, Rabbitbrush, Coyote Willow. Unnoticed squawks cohere into Towhee, Grosbeak, Yellow Warbler, the crazed burble of a hidden Yellow-Breasted Chat. As we pay attention, stories emerge out of vagueness, increasing in clarity. This rock, born of continent-wide, Sahara-like dunes, two hundred million years old; this bird, just returned from México, like me, seeking leafy shade, exploding with song, exclaiming about love.

These intentional changes to our consciousness—simple yet profound shifts in how we speak and think, what we choose to pay attention to, and that we *do* choose to pay attention—help us embrace more of the world, understand it more fully, and feel it with greater vibrancy.

And this is, quite literally, what we were born to do. The evolution of our species—from a naked vulnerable biped on the savanna to successful inhabitant of virtually every habitat on the planet—selected for our immense capacity for attentiveness. We were not the fastest, the strongest, or the most agile. But we did have the gifts of highly attuned eyes and ears, and our inquisitive sense of touch combined with the new twist of our developing cleverness, our facility for memory, and our innovative aptitude for passing knowledge on, story by story. Thus, we could adapt and learn without waiting for our genes to change.

We are built to pay attention to the world around us. A sense of kinship is a natural by-product of this evolutionary heritage.

It is well past time to reawaken to our senses, to reactivate our innate skill at attentiveness, our great natural capacity for being *kin*—animate beings of Earth reaching out for connection.

Let's just say it: we need to *love* this world. Natural history opens the door.

NOTES

1. T. L. Fleischner, "Our Deepest Affinity," in *Nature, Love, Medicine: Essays on Wildness and Wellness*, ed. T. L. Fleischner (Salt Lake City, UT: Torrey House Press, 2017), 8.
2. S. R. Sanders, "A Conservationist Manifesto," in *A Conservationist Manifesto* (Bloomington: Indiana University Press, 2009), 214.
3. R. W. Kimmerer, "Heal-All," in *Nature, Love, Medicine: Essays on Wildness and Wellness*, ed. T. L. Fleischner (Salt Lake City, UT: Torrey House Press, 2017), 237.
4. Kimmerer, 238.
5. T. L. Fleischner, "Natural History and the Deep Roots of Resource Management," *Natural Resources Journal* 45 (2005): 1–13.
6. T. L. Fleischner, "The Mindfulness of Natural History," in *The Way of Natural History*, ed. T. L. Fleischner (San Antonio, TX: Trinity University Press, 2011), 10.
7. C. Norment, *Return to Warden's Grove: Science, Desire, and the Lives of Sparrows* (Iowa City: University of Iowa Press, 2008), 1.
8. H. R. Pulliam and N. M. Waser, "Ecological Invariance and the Search for Generality in Ecology," in *The Ecology of Place: Contributions of Place-Based Research to Ecological Understanding*, ed. Billick and M. V. Price (Chicago: University of Chicago Press, 2010), 85.

SETTLED KIN: COMING HOME TO WHERE WE NOW BELONG

Amba Sepie

How do you weave yourself into a story in a place your ancestors were not born to? The struggle to make kin in strange places is forged by those without an accessible concept of cultural roots, as descendants of those who made long journeys to someone else's land. To belong to a place—to really *belong*—is to see the same tracts of land, rocky outcrops, tributaries, and watery curves of ocean and stream through the many eyes of those who have shared the view across countless generations. It is to know what the soil and skin of Earth tastes like in this place and no other. So how do you ask the land to recognize you when you were not taught to recognize her in this way?

For those like myself, there are many reasons such a legacy of connection has become a half-remembered family history. I arrived on Earth already displaced, too long ago to make any claim on a country, let alone a village or river. My forebears can be counted: four, eight, sixteen, thirty-two, sixty-four, and in less than two hundred years of inheritance my lineage already covers the whole world—sometimes with blood, on one side of the sword or the other. This blood runs from rivers so distant from their source as to grant only a proxy "identity" in a world in which I, and others like me, are expected to be distinct and affiliated.

We settled kin, who were born already colonized, have no legitimate claim on dissatisfaction, although I nonetheless find myself strangely disquieted by circumstance. Here, in a place selected by ancestors I can no longer argue with, who—by force or by choice—picked up and left wherever they once knew as *home*. Slaves, too, came to new lands without options, without a ready manner of relating to a new place. Indentured workers, now as then, refugees, migrants, people pushed into new circumstance without the support of kin back home. No one has the same story, the same skin and kin, the same history, and yet a remarkable number of people have the same problem when it comes to finding a home with Earth where we now are.

The way to "fit in" here is to affiliate, to select one thread of descent from the hundreds available (and we do). Choose just one "interesting" ancestral line in order to say *there, that* is where I am from, and name a tribe, a culture, perhaps a religion. This can lead us in both ad hoc and inappropriate directions. Too often, we remain ignorant of the proper ways to behave in relation to the land here and *her* human kin. A select or stolen identity is never satisfying. I don't want to "belong" to Ireland, or Sweden, or Fiji, and I know that even as I write this, the pull to identify me and name me to a place is so strong that you, the reader, feels that these words have revealed something about me. Simply because I named some places—I wrote it down. I affiliated. And yet, I do not live in any of these places. Nor have I ever. Science tells me Ashkenazi and Melanesian DNA run in my veins, but my skin is pale, my religion is Earth, my location is Aotearoa New Zealand.

What we have in common is that all those who crossed the waters lived thereafter in the wake of having left their old people behind.[1] Not only did they come alone, somewhere between youth and middle age, but all that their home culture helped them bear was eroded by the act of leaving—a history washed away at first by lack of proximity and then by the simple passage of time. *Ocean* appears as country of origin in the census records of some early Americans, in tribal records, too, clearly taken down at a time when just enough forgetting had occurred for no one to be quite sure where Grandfather came from.

For those who survived the dislocation, the land they encountered was always going to be a stranger. In my own story, there are some I can name who were soldiers of a kind—pioneers, to be precise, which derives from the Latin *pedo*, so foot soldier. This is a fitting word for the colonizing acts of those who had been colonized themselves, if long ago. Pioneers, soldiering forth, both inadvertently and deliberately eliminate those communities they come into contact with. I cannot undo those facts of destruction. However, I can point to the weight of history that shaped those individuals into soldiers of empire and love those ancestors too, with the knowledge that they acted according to what they knew. I cannot emphasize enough that the colonized person, just a few generations on from the act perpetrated upon them, unwittingly continues to colonize, for that is what has replaced an emplaced and connected culture.

A culture can dissolve in a mere two generations once the ties to homeland have been cut and the old ways forgotten. Yet this is a key moment in the story—we might assume that people embedded in places willingly took to the seas and away from what they knew. In some instances, yes, but without evidence that there was great reason to leave, the act suggests that the initial dislocation from place was actually much further removed by time. Many who migrated in the movement we refer to as settler colonialism were already thoroughly displaced from the bindings that held their people in place at the point of origin. The cultures of the British Isles and much of Europe had been chipped away at for many hundreds of years, well before the great migrations, with some more eroded than others. I shudder to think how far back we would need to go to find the beginning of the tale. For when you consider how people feel about their homes when they live with a connected or embedded sense of belonging to place, it is utterly evident that no one who lives this way will ever leave by choice. It takes considerable force to uproot a community of people from places that are revered as kin.

There are historically observable outcomes that follow on from place-based dislocations and reasons the pattern of colonization follows a particular set of protocols. To successfully colonize a people, you have to destroy their relationship with their ancestors, human and otherwise, their old people, their children, their land, their animal and sacred kin, their rituals and ways of being. In the end, you have to remove them completely from the places that hold them as kin. This is evident if you examine the scars on the Indigenous communities of the world—the residential school does not do the work alone but quietly dismantles a people through language change, ritual undoing, urban relocation, work programs, food substitution, and westernized health models, among other more nefarious acts. What it means to be colonized is to be dislocated from living in a traditional Earth-oriented way, and this ultimately requires violence to be successful. Observable systemic destruction, as wrought upon the indigenous communities colonized in the past five hundred years of carnage, has happened again and again throughout history, and on all continents, with the difference being only that its most recent incarnation has not been absorbed and forgotten.

When the Haudenosaunee took their case to the United Nations in the 1970s, they stated quite clearly that the major affliction upon the westernized people of the world was that they did not remember how they came to be what they are.[2] And with some rare exceptions, I would argue that colonization processes were already well forgotten before the people of greater Europe took to the seas, before they traveled. The great myths of the British Isles refer clearly to a time when knowledge of the Mother fell to the warrior-kings, when the dragons were slayed and the well maidens raped. The annals of history record the final death blows of the Inquisition across Europe and tell of great hunger and disease as people became divorced entirely from places that once sustained them.

This may have been long ago, but there appears to me some memory, or a yearning, present in all travelers that remains deeply aware of another way of being in relation. This craving can propel those of us caught in westernized nets to seek guidance from those people and communities that appear healthy, whole, and connected. In our ignorance, we can mistake the content of someone else's culture for that sense of connection we really seek. No other culture or ancestors can attend to the strong desire we might have to tap into something *old*, something real, from Earth herself. In truth, culture is a kind of local symbolic architecture that groups of humans employ to sustain their relationships to a place and connected community. It has absolutely no relevance outside the historical context of that group. While such architecture serves to amplify the relationship with Earth, many mistake this for the relationship itself.

There is a way, however, to navigate toward Earth and enter a relationship directly. Earth is our origin and Mother, and the value system, or sentient Earth ethic, that we attach to her is the overriding key to our restoration. She is known in place-based communities to be alive, fully conscious, fully aware, and absolutely in control of herself, her offspring, and her body. Her male aspect who she contains has been often named by humans as the most powerful player in the story, but without his consort he offers us a limited view. She is not absent from anyone, anywhere, ever. Our own connection must be made anew, so that we care for Earth because we *know* she is our origin, not just in some intellectual sense but in the deeply felt sense of understanding we are kin—I am she, and she is me. Such knowledge does not come to the wanderer without effort. If it does not come from generations of blood and sweat in one place, then it has to be cultivated and woven into place where we stand—where we birth *now*.

This can be particularly difficult for city dwellers. Many people may not immediately consider urban places lovable, nor do we extend this to temporary homes.

I'm just here in the meantime, we think.

Awaiting what? The perfect farmlet, the cottage, a croft, a place in the woods? Yet we are called to recognize Earth wherever we are, without confining her only to the wild places, within wild animals, or aspects of ourselves. It is quite irrelevant to Earth whether I, or my ancestors, were born on the precise stretch of limb that I presently occupy, for her whole body remembers me. We are the children in this story, her offspring—and she will embrace us wherever we land, irrespective of the delusions we might hold about the importance of identity.

Identity, conferred by humans and histories, has come to resemble mere genealogical scratchings cobbled together and amounting to lists of names, recent stories, half-remembered dates, and a handful of photographs and old objects that will risk being meaningless two generations on. A place confers identity, for humans or for other animals. In other words, we are named by Earth as belonging where we are *now*. Identity is not a possession or something *we* confer. Earth cares not what we name her—do it as you please, but know that it is *she* who confers identity upon us as kin. "Who am I?" was never a question that could be fully answered by reference to identities conferred by myself or other humans, or by tracing where those many threads of known ancestry lead.

Where I am *now* is Aotearoa, or Te Wai Pounamu, to be precise. The South Island of New Zealand. This is a place where the call to settlers to make home differs a little from how it might be in other parts of Earth's body. Here I am invited to learn the language, *te reo*, and speak with it in formal and prescribed ways. I am encouraged to teach within a bicultural framework, deferent to local Indigenous ways of being, as a part of the response of *tangata Tiriti* (people who came after the Treaty) to *tangata Whenua* (people of the Land). These acts, coupled with the accident (?) of my birth here make this home. It is a forested, often wet, beach-encircled, rocky land carved up into the pastured flat fields of settler farmers; *maunga*, or mountains, pitching into craggy volcanoes; and always *wai*, or water. Here I am, and yet, as I look outward from the shoreline, I am also aware that my ancestral place of origin is right there. *Ocean*.

In this place, perhaps strangely for those from other lands, the traveler and resident alike are called to speak an introduction, in the indigenous language, in order to properly locate themselves. Such greetings were once found in every tongue, naming the connections that sustain us. They bring time, space, and identity together in the unique combination of elements that makes me this person and no other.

There is something powerful in this tradition that unveils what it is to live in connection, and it might sound something like this:

> *I am from the place under this mountain because he provides us with shelter, with guidance, with special plants that grow only there, with somewhere to take our young people to train them. He catches the rain for us and moves it into the waters below. He listens when we ask questions of him, and it is to him that we offer payments to keep the balance between us, and Earth herself.*

> *I am from the place by this river because this river knows me, I was born by her in the same summer the weather was noted for changing early, because she feeds us, because she cools us, because she soothes us in ways only known by water who is teeming with life.*

> *I am called thus, my name, because I come from generations of men and women who lived and gave birth to result in me, here, now, and I honor those named when I speak of who I am.*

This is not a particular Indigenous greeting, but it communicates something of what it means to live in obligation to place. I can say something vaguely like this, but the absence of content reveals how much relationship I am yet to build. These greetings are not commonplace for those who live in this time. Many of us newcomers are not Indigenous, and in truth, we are barely cultured, but there may be generations to come who could be entirely claimed by land in this way.

The core values common across numerous cultures—the values that support our collective obligations to Earth—privilege relationship, respect, reverence, responsibility, redistribution, and reciprocity. There is no mandate of property within these values. You can find them on every continent, though sometimes locked into historical accounts of "how we used to be." They are, however, tightly knitted into the sense of reverence, kinship, and gratitude experienced in traditional Earth-oriented communities and Indigenous ways of being.

Although I came from ocean and walk new shores, I can do this relationship work right here. The journey toward becoming kin and reembedding, restoring, and healing alongside those communities who have suffered more violence than I can fathom or recall—it seems possible. It matters not if my name is known, or that I was named at all, but it matters profoundly that I live in a manner that honors Earth as my origin, that I accept whatever identity is bestowed upon me in a specific place, at a specific time, and then show up for the work that accompanies it. Such work is undertaken by those of us in the here and now on behalf of those who are yet to arrive—inheritors of these obligations, should we choose to instruct them thus.

Today, in this place I call home, *whenua*—land—it is summer. The rain is coming again. She is green, everywhere. All I hear is birdsong. I rest. She inhales, I exhale, we breathe together.

Ātaahua tawhito.

Blessings to you my ancient mother.

NOTES

1. Stephen Jenkinson, author and founder of the Orphan Wisdom School, writes and talks of the relationship between Elders and these troubled histories in Stephen Jenkinson, *Come of Age: The Case for Elderhood in a Time of Trouble* (Berkeley, CA: North Atlantic Books, 2018).
2. Summary in John Mohawk, ed., *Basic Call to Consciousness* (1978; New York: Akwesasne Notes/Book Publishing Company, 2005).

AN ETHIC OF KINSHIP
Kyle Whyte

Evidence from biologists confirms what we witness in the fields and forests around us: degradation of the land and dramatic losses in the populations of species with whom we share the world. The philosophy and practices in the realm of western conservation have not been sufficient to stem the losses. As an Indigenous philosopher, I am drawn to the question of interspecies ethics that underpin these practices, and I would point to an alternative framework based in kinship. Our Anishinaabe intellectual traditions embrace kinship as one way of understanding ethical conduct between species who live together in ecosystems. I want to think about what it means to practice kinship: what are the behaviors that we recognize as making kin?

There are many ideas that other Anishinaabe thinkers besides me have shared regarding kinship. Michael Witgen describes the historical social coordination afforded by human kinship networks among Anishinaabe people who lived several centuries ago. Kinship operated as a "web" of relationships that enabled agreements. He writes: "Real and fictive kinship, established through trade, language, and intermarriage, intersected and crisscrossed over a vast space. These ties made it possible to hunt, fish, and harvest rice, corn, and sugar."[1] Kinship thus enabled agreements through relationships across a cultural and ecological landscape that provided subsistence.

Writing of contemporary relationships with more-than-human beings, the Anishinaabe writer Robin Kimmerer also invokes kinship as the foundation for respectful harvesting in the protocol of the Honorable Harvest, which negotiates the ethics of consuming beings who are also viewed as relatives. She also proposes a new pronoun for relationship with other species that recognizes them as beings rather than natural resources. She suggests we "make that new pronoun 'kin.'" Kimmerer discusses how the kinship pronoun allows us to "now refer to birds and trees not as things, but as our earthly relatives. On a crisp October morning we can look up at the geese and say, 'Look, kin are flying south for the winter. Come back soon.'"[2]

Kimmerer writes further: "Kin are ripening in the fields; kin are nesting under the eaves; kin are flying south for the winter. Our words can be an antidote to human exceptionalism, to unthinking exploitation, an antidote to loneliness, an opening to kinship. If words can make the world, can these two little sounds [*ki* and *kin*] call back the grammar of animacy that was scrubbed from the mouths of children at Carlisle?"[3]

I'm going to follow up on these thoughts about kinship, discussing a few of the lessons I've learned from my own work as a Potawatomi relative, professor, and organizer. Potawatomi is one of the Anishinaabe Nations, and I'll be in dialogue with ideas from other Anishinaabe folks, and will draw upon an array of ideas from Indigenous folks from diverse intellectual traditions.

Kinship Behaviors

From a high-altitude perspective, kinship means a family-like way of defining some of the bonds that connect diverse entities to one another. Oftentimes, kinship is understood as the relationship of blood relatives, but to be kin, one doesn't have to be biologically related. Feelings of kinship can extend to all spheres of life, from parents to friends, coworkers, political allies, strangers, and spirits. Kin can be entities of all animacies: fishes, insects, animals, fungi, plants, water, rocks, or forests. In my way of thinking, kin are any entities that ought to be treated as relatives. What does being a relative entail?

Responsibility is one type of relationship we can have with relatives. While I model my understanding of responsibility based on genetic family relationships, my points about kinship will eventually take us beyond the family sphere. For example, as a family member, I have responsibilities to support my relatives' needs and values, tied to their health, safety, social well-being, economic stability, and aspirations. My relatives, in turn, have those same responsibilities to me. They're mutual responsibilities. It's not only relationships of shared ancestry that involve relatives. Supporting well-being is a responsibility we have to many others in our social circles, environments, and jobs.

In family-like relationships, there need not be contracts or legal documents to define those responsibilities. In fact, when there are, such as in the case of a will, most people I know would say that while those legal relationships may be a necessary part of family life, they are not by themselves kinship. People don't like to reduce the significance of their relationships to the transactional dimension of a will or contract. We intuitively know that relationship arises from the emotional realm of mutual care.

Kinship relationships within a family provide an important layer of support for individuals. In an ideal situation, exercise of mutual responsibility to various family members can provide a web of support and care, both physical and emotional, that can sustain individuals through the very toughest times.

When those who are expected to be kin do not treat one another as relatives, such as exploitation in the execution of a will, then the effects can be devastating. At the same time, when kin relationships are strong, something like a will can be executed with great care. Such care exceeds the transactional dimension of the will, drawing on enduring kinship bonds as people deal emotionally with the passing—the walking on—of a loved one. Bonds of respect, affection, and care for a relative's well-being transcend any performative obligation.

Now, imagine how such kinship would be practiced among relatives of different species and diverse animacies, like plants, animals, or water. It's one thing to protect these relatives through regulatory laws, policies, or conservation easements. Some laws or contracts may be intended for perpetuity, but they can be overridden if a different political party comes to power or if people are in conflict over land use. It's quite another thing to protect them as an ethical responsibility. As relatives, humans have to acknowledge that they depend on other beings such as plants, animals, and water. There's mutual responsibility among them. A change in a law or a broken contract does not change that responsibility. In Indigenous philosophy, being a good relative is an ethical and spiritual responsibility. An ethical responsibility of kinship has greater longevity than a legal obligation.

This concept is illustrated in the writing of Salish traditions, such as a story Lee Maracle shares about the responsibilities connecting humans and salmon:

> Sockeye were sent to Salish women to assist us during times of hunger. We were asked to honor sockeye and take care of the waters. We were told that if we take the sockeye or their habitat or the women for granted, they would not return. The story does not say that if we lose our fishing rights, we are not responsible for caretaking the fish or the women. It does not say that if we allow the newcomers to desecrate the waters, we are relieved of responsibility. It says that if we don't take care, they will not return.[4]

Kinship and Society

Kinship plays a pivotal role in societal life at many levels. The scale of kinship ranges widely, from families or small social circles to collectives such as nations or peoples. The cultural agreements of a society have an impact on how its members can live, whether they can experience freedom, justice, love, and inspiration. Social fabrics also have to constantly withstand the disruptions of a complex, constantly changing world. The spread of disease vectors, sea level rise, and economic insecurity are related to complex factors tied to climate, culture, weather, capitalism, pollution, and group psychology, which can both affect and be influenced by kinship relations.

Kinship relations can also contribute to societal responsiveness in a constantly changing world. Societies can anticipate and prepare for some changes. Preparations for ordinary seasonal changes can be anticipated, such as the seasonal transitions in the Great Lakes region, which require collective action for adaptation. Hurricanes or earthquakes in other regions may not be fully predictable, yet decisions about infrastructure and emergency management can be made in advance. The interrelated responsibilities of kinship in society allow collective action that creates resilience in the face of change.

The longevity of a society relies more on adaptability than stability. This philosophy is firmly rooted in Anishinaabe intellectual traditions. Heidi Stark writes that "the Anishinaabe transformed themselves, adapting to their ever-changing environment: Anishinaabe stories often conveyed the importance of change" and the capacity to respond to it.[5] Gerald Vizenor's concept of survivance refers to "the continuance of native stories, not a mere reaction, or a survivable name. Native survivance stories are renunciations of dominance, tragedy, and victimry."[6]

I've often used the concept of collective continuance to refer to the resiliency of a society's fabric. Resiliency is not just about bouncing back. Rather, it is about a society furnishing a support and empowerment network for the sake of its members' cultural integrity, health, economic vitality, and political peace. A high degree of collective continuance means that societal life is organized in ways that can anticipate potential changes, learn from the past, and mobilize everyone to respond when called for.

Collective continuance is about kinship. Kinship bonds serve as models for how members of societies should treat one another in easy and more challenging times alike. That is, kinship relationships can generate high degrees of collective continuance. Consider how family members band together to address unexpected crises. Now imagine scaling that up more broadly to an entire society characterized by mutual responsibility, within human society and in the wider web of relationships with other species.

Chie Sakakibara's book *Whale Snow* is a multispecies ethnography about mutual responsibility between Iñupiat and bowhead whales at the societal scale. *Cetaceousness* means the kinship whereby humans and whales act together and through each other through the types of bonds I'm calling kinship. The relationship is one way to assess the risks of climate change and maintaining resiliency for future generations. Regarding cetaceousness, Sakakibara writes that it's a "force with which to confront future adversity."[7]

Qualities of Kinship

Kinship qualities are crucial for different entities being able to band together across diverse animacies, whether in families or in societies. What are the qualities of kinship relationships that generate care and responsiveness? There are many such qualities. Kinship is complex. I'd like to discuss just a few of those enabling qualities: responsibility, consent, and reciprocity.

To begin with, responsibility is a type of kinship relationship. To really understand kinship though, we need to look further at the qualities of relationships that involve responsibilities. The mere acknowledgment that I have a responsibility is of little value to me or anyone else. I have to exercise the responsibility through certain qualities.

In a kinship framework, responsibilities should have the quality of consent. Consent is a quality of responsibility by which relatives have the utmost accountability for honoring one another's freedom. To respect someone's consent is to respect that person's independence, bodily security, and self-determined choices. When times get tough, I hope I have relatives who won't take advantage of my vulnerabilities or make decisions without consulting me. Responsibilities must be exercised consensually.

Between humans, it is possible to aim for genuine respect for consent. But what about consent between humans and insects, or between humans and the climate system? Humans may not be in the position to be fully confident that they are treating plant, animal, and fungal relatives consensually. Some animacies are hard to read for humans.

But that does not mean that their consent is not sacred. It is. The Anishinaabe writer Basil Johnston recounts a story about humans abusing the generosity of animals, violating animals' consent by treating them as exploited servants. The animals had to have their own council to discuss their oppression by humans. The story highlights why we must always concern ourselves with whether we are being respectful of the consent of animate entities who may not be able to communicate as directly with us as other humans can.

But it can also be the other way around. Animals, weather, and others do take actions for which humans do not give consent. Melissa Nelson describes Anishinaabe stories about Mishipizhu and the Thunderers as they relate to water. These powers are in tension, as Mishipizhu is a "guardian" and "keeper of balance," and the Thunderers represent, among other things, extreme water events. As climate change occurs, the tension erupts into extremes of too much or too little water during different periods that are harmful to humans and other relatives—harms we don't consent to.[8]

Yet Nelson asks us to ponder why these things happen. Have humans done something wrong to upset this balance, and what can they do differently? Should humans be more respectful of the tension between Mishipizhu and the Thunderers? Humans' initial lack of consent subsequently provokes respectful thinking and planning about how to be responsive to forces that are outside our control while respecting those forces' autonomy.

Reciprocity is another quality of relationships, one in which each relative is confident that the investment made in the well-being of others will be gifted back to that same relative. Reciprocity is a quality marked, at least for humans, by having an emotional sense of mutuality. When we strive to give gifts to others, we have a reciprocal relationship when we know our gift is meaningful.

In an essay I coauthored with Nick Reo, we wrote about the ethics of Ojibwe white-tailed deer hunting. Some of the hunters we interviewed expressed how profoundly they wanted to honor the gift of nutrition that deer gave to them. They stated the importance at an emotional level of ceremonies and demonstrations of respect to show the deer that humans are grateful. Such actions motivate humans to take greater care of deer habitat, bringing reciprocity full circle.[9]

Imagine a society in which consent and reciprocity are richly practiced as qualities of mutual responsibilities spanning diverse animate entities. In such a society, humans show high regard for the consent status and reciprocity status of all their relationships. When times are tough, don't we want to turn to those with whom we have consensual and reciprocal relationships? We know we can count on them because they respect our consent and reciprocity, as we respect theirs.

Can we expand our imagination of kinship to include our relationships with plants, animals, and other animate entities? If we take seriously consent and reciprocity with plants, wouldn't that motivate us to spend more time with them? And wouldn't we be encouraged to learn about them in order to understand what consent and reciprocity might require? Wouldn't that time and learning lead us to be more aware of the negative impacts plants are facing, and also prepare us with a more ecological understanding of what we could do to protect them as we exercise genuine consent and reciprocity?

While simple, my final questions suggest the possibility of a society with multiple spheres of kinship that go beyond just our closest family and friends. The level of intimacy varies across relationships, from plants to human neighbors, but we can certainly imagine and practice social relationships where our work, political, religious, and environmental responsibilities embody the qualities of consent and reciprocity—not only with our fellow humans but also with the diverse animacies with whom we share the land.

NOTES

1. Michael Witgen, *An Infinity of Nations: How the Native New World Shaped Early North America* (Philadelphia: University of Pennsylvania Press, 2011).
2. Robin Wall Kimmerer, "Nature Needs a New Pronoun: To Stop the Age of Extinction, Let's Start by Ditching 'It,'" *Yes!*, March 30, 2015, https://www.yesmagazine.org/issue/together-earth/2015/03/30/alternative-grammar-a-new-language-of-kinship/.
3. Robin Kimmerer, "Speaking of Nature," *Orion*, June 12, 2017, https://orionmagazine.org/article/speaking-of-nature/.
4. Lee Maracle, *Memory Serves: Oratories* (Edmonton, AB: NeWest Press, 2015).
5. Heidi Kiiwetinepinesiik Stark, "Marked by Fire: Anishinaabe Articulations of Nationhood in Treaty Making with the United States and Canada," *American Indian Quarterly* 36, no. 2 (2012): 119–49, at 124.
6. Gerald Vizenor, *Manifest Manners: Postindian Warriors of Survivance* (Middletown, CT: Wesleyan University Press, 1994).
7. Chie Sakakibara, *Whale Snow: Iñupiat, Climate Change, and Multispecies Resilience in Arctic Alaska* (Tucson: University of Arizona Press, 2020).
8. Melissa K. Nelson, "The Hydromythology of the Anishinaabeg: Will Mishipizhu Survive Climate Change, or Is He Creating It?" in *Centering Anishinaabeg Studies: Understanding the World through Stories*, ed. Jill Doerfler, Heidi Kiiwetinepinesiik Stark, and Niigaanwewidam James Sinclair (East Lansing: Michigan State University Press, 2013), 213–33.
9. Nicholas James Reo and Kyle Powys Whyte, "Hunting and Morality as Elements of Traditional Ecological Knowledge," *Human Ecology* 40, no. 1 (2012): 15–27.

KINSHIP WITH PLANTS
Matt Hall

Place

Our family house sits on a range of hills that extends toward a small bay within the Taputeranga Marine Reserve, at the southernmost tip of Aotearoa New Zealand's North Island. Behind our house, the slope rises steeply. We have a small lawn area surrounded by a herbaceous garden bed and some feijoa trees (*Acca sellowiana*) that we planted when we first moved in. Above are a couple of terraces, first with the washing line, a young plum tree, and a small garden shed; then some more vegetable beds and fruit trees, including a majestic and bountiful walnut (*Juglans regia*). Behind that, the remainder of the section is native bushland, regenerating after widespread forest clearances performed during colonization.

Aotearoa is a special place, known globally for recent recognitions of rivers and forests as legal persons. Māori recognize and celebrate *whakapapa* (genealogical ancestry) that directly binds humans and nonhumans, including plants, in a web of intertwining kinship connections. The Māori creation story of the Sky Father (Ranginui) and the Earth Mother (Papatūānuku)—and their children, including the forest ancestor Tāne—recognizes a genealogy in which all beings share a basic common ancestry. This worldview is "at the heart of Māori culture, touching, interacting with and strongly influencing every aspect of it."[1]

Our family is predominantly Pākeha, or what the New Zealand census calls "New Zealand European." A Pākeha in Aotearoa, I find myself somewhat dislocated and disconnected. Despite a clear genealogical link with other living beings, which science has acknowledged since Darwin, I do not have a living cultural tradition that recognizes kinship with plants. My ancestors helped originate a modern industrial culture that predominantly views the plants with whom we live as passive, insentient beings, radically different from humans. In this age of destruction, I spend time in the garden wondering how people like me can begin to recover a sense of kinship with plants. I invite you to head into your garden, whether private or public, to think about this along with me.

Kinship

My understanding of kinship is influenced by Marshall Sahlins, for whom kinship is a "mutuality of being" in which "kinsmen are persons who belong to one another, who are members of one another, who are co-present in each other, whose lives are joined and interdependent."[2]

Such mutuality of being emerges from either a shared ancestry or the creation of family relationships through marriage, with this latter, *affinal* kinship being most common. Knowing then that a scientific recognition of genealogy is not enough, how can we build such relations of affinity with plants so that we recognize our shared lives and interdependencies?

In our garden, I've been thinking about a number of strategies for building affinity that are both personal and scalable, and that, if promulgated, could help to reseed a culture of human-plant connection. I think of them as a tentative first step toward much needed sociolegal reform. Like any good taxonomist (and I'm a pretty bad taxonomist), I've put these into nice neat categories, but of course, the living world is far more entangled and interesting than that.

Deep Observation

Thomas Lowe Fleischner has advocated for natural history as "a practice of intentional, focused attentiveness and receptivity to the more-than-human world, guided by honesty and accuracy."[3] Through this deep observation, Fleischner has argued that through natural history as a practice of kinship we create "a forum for interaction with Others, encouraging compassion and respect, helping us rediscover passion for the world and each other."[4]

I have a great deal of sympathy for such a deep-observation approach to other living beings, stemming to a large extent from my own background as a botanical scientist. In the garden, I take as much time as I'm able to observe what the plants around me are doing and what condition they are in—whether they are diseased, whether they require watering or feeding, whether they are beginning to fruit or flower. When my children come out in the garden with me, I try to point out basic plant anatomy, but I also explain how variable these parts can be in plants. The idea is to get them first and foremost to spend time with plants and to look at their forms, but also to understand some of the basic patterns and functions in the diversity of the life around them. I also try to help them say the names of the plants, both the common English names and the Māori names, if I know them. On some occasions, our eldest daughter tells me a Māori story that she has learned at school. Her favorite is a tale of Tāwhirimātea, the *atua* of weather, who, in anger at his parents Rangi and Papa being pulled apart, throws his eyes up to the heavens, where they become the stars.

Building this familiarity is a first step to affinity, but I do often wonder whether it is sufficient. Our cultural framing of plants as passive, insensitive, inferior beings is not easily overturned by attentiveness to their forms, ailments, or names. There are several reasons for this, including the fact that our utter reliance on using plants for our survival means that it can be difficult to break out of our predominantly use-based relationship, and, for human beings, plant behavior is often imperceptible to the naked eye. In European-descended cultures, we have also socially and legally constructed plants as objects, mere property for human use. This makes it difficult to build the links of affinity. In my own practice,

I find that observation is fundamental but does not bring me to a feeling of kinship with the plants in our garden. The same was true when I worked as a botanist and spent large amounts of time observing plants both in the field and in the herbarium.

However, observation does play the role of slowing down my mind and allowing me to come into the presence of, and be present with, the living beings all around me. As a supplementary practice, when in the garden, I extend observation into a meditation on the raw physical connections between my body and those of the plants around me. For example, take the old walnut tree that provides a bounteous supply of walnuts each autumn. I mentally rehearse how the flesh from the walnut both nourishes and becomes my body, the lipids and amino acids in the walnuts breaking down to form my own flesh and bone. If we practiced the ancient practice of night soil, then I would be able to imagine the completion of the nutrient loop.

However, I do also mentally conjure our affinity through shared breath, forming mental pictures of the passage of oxygen from plants into my body and the carbon dioxide exhaled from my body into the body of the walnut tree. I find that a simple exercise like this forms mental and emotional connections and reduces the vast ontological distance between us (at least for me). Every time I do it, this practice reminds me both of the idea of "mutuality of being" and the ancient stories of metamorphosis between plants and humans, one of the key themes in the myths about plants from cultures that do recognize kinship, such as the Gunwinggu tale of Namalbi and Ngalmadbi, a husband and wife who were transformed into pandanus trees.[5]

Personal Interaction

Elsewhere, I've defended the concept of plant personhood and the acts that animists do to enact, create, and recognize that personhood. I would like to take this a step further and point out that all the activities described in this essay are ways of recognizing personhood in plants. Such recognition of personhood is fundamental to Sahlins's description of kinship.

Although animists do not divorce use from personhood, here I emphasize a relationship with plants beyond use, because the idea that "plants are only there to be used," which dominates so many contemporary cultural understandings of them. If we think about common interactions with plants—observing their beauty, caring for them by watering or feeding, picking their leaves, eating their roots or fruits, cutting them down for wood or timber—many involve some kind of instrumental use. In such engagements, a beautiful rose is predominantly an object of beauty, rather than another living subject, and a new vegetable garden is lovingly cared for because it will provide tasty food at the end of the season.

In our desperate ecological situation, engaging in habitat restoration is a pragmatic form of noninstrumental interaction with plants. In dominating the plant kingdom, and placing plants at the bottom of a human-invented hierarchy of worth, we justify a radical and almost total transformation of the Earth's natural plant habitats into sites for human use. The act of mindfully giving back land to plants via restoration is a direct acknowledgment and reversal of this domination. If pursued in this way, it is a powerful demonstration of kinship.

Over the past five years, my wife and I have grown native seedlings of the *ngaio* and *taupata* and replanted these in our garden. We planted these seedlings in large open areas being smothered with invasive weeds, and in the intervening period they have grown into juvenile trees over two meters high, suppressing the weeds and providing more habitat for native birds such as tui and silvereye. In other areas of the garden, we have simply let the weeds flourish, to create a haven for insects and more food sources for insects and birds. Unfortunately, this is not a "one and done" type of endeavor but more a life's work. Each time I go into the garden, I take time to observe these areas in which we have slowly begun to relinquish dominance of the plant kingdom, and with each observation I try to wear away my psychological illusion of superiority and control. I have had most success in observing the speed and force with which the new *ngaio* trees grow. Each time I take time to observe their thickening, corky trunks and proliferation of yellow-green lanceolate leaves, I become more attuned to their power and presence, their vitality and activity, and less concerned with their origins as cultivated seedlings.

Ritual

I base my understanding of ritual on Ronald Grimes's seminal work. For Grimes, ritual is a type of performance, and what he calls the "currency" of the gift economy. Unlike the "real" economy, the gift economy "originates with a give-away, a proffering of gratitude magnanimous, of play excessive and impractical. . . . [I]t assumes the necessity of loss, even of deliberate and celebrated loss, of sacrifice, of giving up what you'd rather keep."[6]

These ritualistic performances are oriented toward those beings (plants, animals, and rocks, among others) with whom we share the Earth, and who make our life on Earth possible. We need them in order to move from mere words into action, and to directly undercut the foundations of the mess that we have made of this Earth. The thing that humans have most to lose, and the thing that Grimes highlights humans most need to give up, "is their false sense of themselves as superior."[7]

If this idea of ritual seems too grandiose, far-fetched, or perhaps disconnected from the "real" work that needs to be done, Grimes guides us away from the global scale to ritualizing in our own backyards. In our backyards, he asks us to imagine "a single gesture or posture that might become the seed of a rite. Make it one worth doing, or holding, over and over. This is ritual, so again, again, and again. Even when no one is watching: again."[8]

Our guide is imagining a gesture "so simple and profound"

that, even if it doesn't "save the planet," we would keep doing it in the hope of entering the gift economy.[9] For me, that is personally reinhabiting our world on kinship terms rather than as the ecological tyrants we have become.

Since I first read Grimes's article, I have played with quite a few ritual gestures toward plants, from kneeling beneath trees to prostrating myself in deep grassland. I have played with Robin Wall Kimmerer's suggestion of calling our kin *ki* rather than *it*, but for some reason that word does not quite resonate.[10] In our garden, I am left with two simple rituals.

The first of these involves the simple acts of asking permission and offering thanks, when I take something from a plant in the garden, be it the leaves of an herb, walnuts from the ground, or a flax leaf to use as garden twine. Yesterday, in the evening twilight, I went to get some basil mint (I would love to grow basil, but it withers in the Wellington winds) for a dinnertime dish. As I bent down over the low, brick wall that comprises the bed edge, I said a quiet "hello," followed by a short pause, then a soft "thank you" once I had plucked a few leaves. As I went back inside, I carried part of a living being with me.

If I am honest, sometimes it feels strange, sometimes forced, and sometimes I forget. I am most comfortable doing it when nobody is watching, I think because it feels deeply personal and extremely humbling. It is so simple, but that one gesture completely reorientates my brain from just grabbing a bunch of leaves to realizing that, without these beings, I am nothing. This humility opens a space for sympathy. It also breeds gratitude. I would like to tell you that I maintain this throughout the day, but I do not. So back I go to thank the basil mint and the walnut tree again and again and again.

My second ritual is more of a posture than a gesture, and a simple one at that (thanks Ronald!). Whenever I get to go out into the garden (and with two small children, that is not always as much as I would like), I sit and listen. I listen in the way the Aboriginal Elder and philosopher Bill Neidjie has taught us to listen:

> That tree now, feeling . . .
> e blow . . .
> sit quiet you speaking . . .
> that tree now e speak . . .
> that wind e blow . . .
> e can listen.[11]

By listening I am not trying to discern audible sounds from the plants. Instead, I am hanging on Bill Neidjie's line "that tree now e speak" and using it to guide me. I sit on the edge of the vegetable bed underneath a *ngaio* tree and I listen. As I listen, I am trying to step back, to stop my own verbal commentary, to put aside my own human wants and desires and to allow the plants to take their turn at describing, shaping, and living in the world. It is akin to a botanical meditation, with space for other beings' flourishing as its object. Sometimes I feel like crying; other times I do not feel much. We have been dominating plants for thousands of years, so this is bound to take a while.

Cultivating Sympathy

My understanding of sympathy is heavily influenced by Max Scheler, who, writing in the 1920s, gave sympathy a central role in his philosophy. Scheler's understanding of the nature of sympathy was a visceral, embodied act rather than a projection or reproduction of our own human experience. The term *sympathy* covered a range of diverse concepts such as fellow feeling, empathy, and pity. The philosopher Dan Zahavi elegantly breaks this down for us: "More basic than sympathy is what Scheler terms Nachfühlen, which I . . . render as empathy. In short, whereas empathy has to do with a basic understanding of expressive others, sympathy adds care or concern for the other."[12]

In many ways, these two aspects—developing a basic understanding of other beings, and a care or concern for them—are threaded throughout the activities that I have been discussing in this short essay. Engaging in such acts, then, can also be viewed as a way of deepening an empathetic affinity with plants and cultivating concern for their welfare, both of which are key components of kinship relationships.

In other ways this sympathy is foundational to a project that seeks to recover and restore kinship relationships. Purposefully pursuing a philosophical ecological restoration or jettisoning human claims to exhaustive use of "natural resources" makes sense only when it is grounded in some existing care for their welfare. There must be some preexisting care or concern underpinning this will to connect. Without that, I struggle to envisage why else someone would act.

We drop down, then, to a foundation of empathy, a base understanding of expressive others. Fleischner may be right that a practice such as natural history or observation can be a key component in this cultivation of understanding. Such understanding should include contemporary scientific understanding of plants' remarkable sentient capacities. To propel us into caring, we could do worse than to head out into the garden and contemplate the myriad ways these sentient lives are intertwined with our own.

NOTES

1. Manuka Henare, "Tapu, Mana, Mauri, Hau, Wairua: A Māori Philosophy of Vitalism and Cosmos," in *Indigenous Traditions and Ecology*, ed. John Grim (Boston: Harvard University Press, 2001), 202.
2. Marshall Sahlins, "What Kinship Is (Part One)," *Journal of the Royal Anthropological Institute* 17, no. 1 (2011): 11, https://doi.org/10.1111/j.1467-9655.2010.01666.x.
3. Thomas Lowe Fleischner, "Natural History as a Practice of Kinship," *Minding Nature* 12, no. 3 (Fall 2019): 12–15, https://www.humansandnature.org/natural-history-as-a-practice-of-kinship.
4. Fleischner.
5. The Gunwinggu (Kunwinjku) people are an Australian Aboriginal people, whose country is in West Arnhem Land, Northern Territory, Australia.
6. Ronald L. Grimes, "Performance Is Currency in the Deep World's Gift Economy: An Incantatory Riff for a Global Medicine Show," *Interdisciplinary Studies in Literature and Environment* 9, no. 1 (2002): 157.
7. Grimes, 157.
8. Grimes, 162.
9. Grimes, 162.
10. Robin Wall Kimmerer, *Braiding Sweetgrass: Indigenous Wisdom, Scientific Knowledge and the Teachings of Plants* (Minneapolis: Milkweed Editions, 2013).
11. Bill Neidjie, *Story about Feeling* (Broome, Australia: Magabala Books, 1998), 18.
12. Dan Zahavi, "Max Scheler," in *History of Continental Philosophy III*, ed. A. Schrift (Edinburgh: Acumen Press, 2010), 178.

MERCY

Nickole Brown

Speak, beast.
    Speak.

  Curled suburban cuddle in front of the fireplace,
  possum-dreaming and paw-kicking, your fur
  a tumbleweed down the waxed hallway, your ears
  cropped and your tail docked,
      speak.

  Or how about you? Yes, you: spotted neck stretched
  towards what's left of your acacia trees, neck as long as
  a man's grave is deep, I need you to fire that impossible
  distance between your heart and tongue and
      speak. Speak,

  you barred owls with the pink tip of a poisoned
  mouse dangling from your beak, all your many neck bones
  hinged in one spot so you can pivot your blinking
  face to me. I am watching; I am waiting;
      look at me and
      speak. Speak,

  you black snake drowsing on the hot blacktop,
  your forked tongue remembering the long kiss
  of voles in tall grass, a memory gone when the woman
  runs over you—then twice again, just to be sure. I need you

to speak. Speak,

you jittering squirrels, you murder of crows, you quarrel
of sparrows, you pitying of turtledoves, you everyday
outside flick of life at the mercy of these coming winds,
these rising
waters filthy and licking with flames.

> Speak, sing to me, caw and fuss
> among what brittle branches
> left, I have opened my windows;
>
> I am listening.
> Speak,

you hellbender—giant salamander you are—a rarity
now put on educational display, you eel-looking
haint once pulled up by fisherman in these mountains.
    Speak, because no one knows who you are
anymore; you must make us remember. Turn your
slack maw to my tapping on the bent plexiglass
    and speak.
    Say what it is you need to say.

Oh, and you. You with your thick skull blown apart
with a high-caliber swagger, your memory long
and your once trumpeting
    gone, speak to me
    from beyond.
    Tell me
what it was to have those ears of yours in fury,
raised like giant gray flags, how hard you fought,
and even once shot how you just stood there
confused, already dead but refusing to fall

   until your knees buckled, the rest of you slumped,
and the great pillars of your tusks were
chain-sawed from your face.
      I need to hear especially
from you: I have a photo here of a man
grinning with your lopped-off tail in his fist.
I need you
  to speak, dammit,
  speak.
  Say what it is you need to say.

Or how about you: kin but safe
in a cage, I've heard your placid
chewing at the zoo—you took a sweet potato
from my hand with the wet, breathing end
of your trunk, slid it into the deep
socket of your mouth.
  I know *that* sound—
I smiled at your keeper, fed you another treat, but now,
  now, I need to you to speak.
  Try it—swallow down
  that food and flex
  your tongue, push out the
  grassy air from those
  miraculous lungs.
  I am waiting;
  we are all waiting.
  I am begging. Please, beast.

  Speak.
  We are running out of
  time; there are so many us, and what's left
of you are
swaying in pens, rocking from side to side, sleeping and aching

and craving and thinking—I *know* you are thinking—
but not saying a damn word.

> No, not one, not one,
> which I all we need,
> I swear—all of you,

you birds and cicadas, all you flying, leaping, vine-grabbing
canopy beings, all you furred and quilled things too,
or you chittering in your burrows and hiding the dark—
step from the mouth of your dens and speak.

> Listen to me,
> you belong here,

you with a mouth full of milk, with dirt under your claws, all of you
rooting and panting and hibernating and ruminating and standing
by the fence, watching us speed past in our cars.

> Listen to me.
> I need you to try,
> try to say
> just one word.

We can start slow with the low pleasure
sound, the delicious *mmmm* that closes the door to *home*,

then growl out that middle vowel and what comes after,
let the back of your throat issue forth a warning
and mean it, get angry with me,
because you've had enough, because if you don't find these sounds

there will be nothing left.

Now, you're almost there—

pull back your lips if you have them, show your teeth
if you have them, hiss, let out all the air, whatever air is left
and in doing so take us back to all our beginnings
with the sound of waves, with that final syllable—*sea*.

> Now try again.
> Try again.
> Try, please try.

> All at once.
> It's such a small
> word on which
> your lives depend.

> I'll do anything.
> Please, beast. Speak.
> Say it, with me, now:

> *MERcy, merCY, mERcy, meeercy, merrrrcy, mercy, mercy, mercy.*

> Now, please. Try. Try again.

> Repeat after me

ENTANGLED KIN: MERGING BIOTIC AND ABIOTIC BEINGS

Ajay Rastogi

The day on which I observed water and roots playfully sharing nutrients was the day I began my journey to understand that all beings, all systems, were kin to one another and to me. In 1988, I was a master's-level researcher in the plant physiology laboratory at the University of Pantnagar in northern India. We were concerned that large-scale eucalyptus plantations in India were leading to phosphorous deficiency in soil. The phosphorous seemed to be locked in the leaves during the decomposition process and not released back into the ecosystem. I sought to scientifically validate this concern by speeding up the decomposition process and reducing it to one component in a controlled experiment.

So I planted eucalyptus saplings in two mediums, sand and water, that had been completely sterilized. I added radioisotopes of phosphorous to the mediums to enable uptake in the leaves. I powdered the leaves and measured the quantity of phosphorous available. We made recommendations on this basis with statistical confidence in our results. But this research did not occur in the forest; this research did not surround me with the invigorating scent of a eucalyptus grove; this research did not plant my feet anywhere near the settlements and communities living close to these trees. This research placed me under the fluorescent lights, working upon the cold slab tables of a laboratory, walking in and out of doors with warnings of radiation hazards.

The experiment was highly mechanical, requiring a highly reductive view of nature as a set of bits and pieces that could be broken down to serve human desires. I was asked to overlook the nonnative tree's devastating impact on native soil and water tables. Beyond that, I was asked to break down the forest ecosystem—the soil, the water, other flora, the people who shape the forest with their livelihood every day—not just to one species, not just to one tree, not just to one leaf, but to one isotope in hopes of maximizing the economic efficiency of a forest system. It seemed as if the ultimate purpose of this experiment, and of the forestry regime and economic system that the forestry regime supported, was to reduce the animate, interconnected, and biotic nature of a living being to an abiotic commodity—there to serve only the human thirst for profit.

But one day, the plants in the experiment revealed something fundamentally more than efficiency to me—they revealed their kinship. The experiment was run in a hydroponic system. In this westernized university, I was taught to see the water as merely a dead carrier of nutrients. Yet on this day, I looked into the squarish sea encased in a plastic tank and marveled at the roots actively interacting with and dancing within the water. I noticed a conversation happening among the roots, the nutrients in the water, and the water itself. I realized that what we called *abiotic* (the water, even the nutrients) and what we called *biotic* (the roots) were entangled as one, in relation, as living kin. The roots, the water, the nutrients, and I were "kinning." Never again could I reduce such a living dance of entanglement, such kinning, to the dichotomy of abiotic and biotic for the sake of efficiency.

Food as a Pathway to Merge Biotic and Abiotic Systems

Not everyone has the privilege of time in a laboratory to come to consciousness about the interconnectivity of living systems. Yet we all have the opportunity to experience the entanglement of biotic and abiotic beings through the food systems that we participate in every day. Recent health concerns around food have accelerated a social revolution of "conscious choice," making my lab-based epiphany about kinship accessible to anyone who seeks more ethical ways to eat.

The realization that food on our table is drained of nutrients and may contain toxins, allergens, and chemicals (which mimic hormones and adversely affect our body's physiology) has driven a revolution into whole foods, organic production, and processing. Across the globe, consumers are calling for heirloom varieties of seeds and crops. Historically, these heirloom varieties were discarded because of concerns about efficiency in production, uniformity, and appearance. It is time to bring them back for better nutrition. Local food's reducing of food miles, plant-based diets' reducing of deforestation for livestock rearing, regenerative agriculture's sequestering of carbon in the soil—this revolution has expanded human care from physical health to planetary health, through, for example, climate actions for and from the farmers around the world who have deep expertise with heirloom varieties.

The climate crisis is also leading to greater realization of the entanglement of the biotic and abiotic—that soil is a living being entangled with the health of everything: the worker in the field, the plants being grown, the diet of the consumer, and the amount of carbon in the atmosphere. Food reminds us that all is connected, and industrial food systems show that we ignore this ecological kinship created through food systems at our peril.

Lessons from the Past for the Future

Many communities are looking back to learn from traditional cultures that shaped biodiverse landscapes to sustain human communities and the overall richness of life. Many of these cultures offer lessons for kinship across the western biotic-abiotic binary. Traditional communities could qualitatively enhance their habitats and keep them fertile, diverse, and clean, despite not being able to quantify their ecosystem services with the same metrics used today. However, their ability to utilize nature's resilience and equilibrium sustainably over thousands of years reflects not just their science but also their practices of the fine art of living. Often the biotic and abiotic were linked through a collective consciousness of "sacredness." Such communities did not think in terms of single components: soil as a medium that needs to be replenished, water as something that needs to be hygienic, air as a thing that needs to be clean, livestock as a resource that needs to be healthy, and forests as vital habitats that maintain the population dynamics of several species. Rather, they understood all these together as an interconnected system with human beings as a part of the ecosystem, as kin.

Indigenous Principles

Although this sense of interconnectedness was lost under an industrial agriculture paradigm, a revival is emerging around zero-budget farming, minimum tillage, cyclic agriculture, and ecological farming. In addition to this reshaping of applied farming alternatives to industrial agriculture, the revival of ceremonies and festivities that celebrate the earth and natural elements surrounding food systems are leading to the resurgence of sacred relationships.

For example, among the Lepcha community in India, spiritual practices, in particular, reinforce relationships with ancestors who previously lived in the same landscape and have passed on associated knowledge and wisdom to current generations.[1] Weaving four layers of social and ecological kinship, the Lepcha particularly emphasize the following principles:[2]

- **Reciprocity:** Equal exchange within society and with the natural and spiritual world
- **Solidarity (or brotherhood or sisterhood):** Unity with the human, natural, and spiritual world (but especially within the community or human world, such as helping those in need)

- **Equilibrium:** Balance (or harmony) within the human, natural, and spiritual world
- **Collectivity:** Oneness of the human world with the natural and spiritual world

How are these principles reflected in day-to-day life?

Lepchas show that reciprocity is reflected in traditional practices of sharing and exchanging seeds and services within the community. This age-old practice has continued for generations, across centuries. Traditional practices of planting new saplings after felling trees and exploring new crops and varieties are embedded in customs of the community that support sustainable uses and conserve agro-biodiversity. In other words, social kinning (sharing and exchanging) leads to ecological kinning (agro-biodiversity).

Second, Lepchas practice solidarity through a community organization called the Sozom, which takes up the issues of conservation through the promotion of language, culture, community rights, identity, and upliftment. It proactively seeks out people in the community who might need help and provides it through contributions from all others in the community. These contributions may not be solely financial but also material, physical, psychic, and social, providing emergency relief and long-term systemic support. In this sense, the work differs from that of conventional charities that provide freebies without upholding mutual respect and enhanced self-reliance. In fact, a Lepcha person does not ordinarily demand help, and the Sozom has to seek individuals out informally through other community members—the kinship network ensures that the help will come.

Equilibrium for Lepchas is maintained through beliefs, rituals, and practices that reinforce that human beings are a part of nature. These rituals strengthen relationships with a landscape regarded as animate: mountains, caves, rivers, water sources, grand old trees, rocks, and farm land. Along with the ancestors, such prominent geographical features are often referred to as the deities in the landscape. Lepchas consider themselves to be the children of Mount Khanchendzonga. Their use of natural resources reflects sustainability in various ways. When they harvest mushrooms, they make sure that their spores are left behind for adequate regeneration in the next season. There are prohibitions on hunting and fishing during the breeding seasons of birds, animals, and fish. Precise decisions are made on when, how, and which trees to cut for fodder depending on the types, needs, and life stages of the livestock. The management of family-owned, community-owned, and wild landscape units is conducted according to customary norms without any fencing and external monitoring. Ecological kinship strengthens social kinship.

Finally, when it comes to collectivity among the Lepchas, collective decisions and abiding faith sustain the self-governance of community regulations. Like many mountain communities, Lepchas see the world in totality. While performing rituals related to nature, they connect themselves with a spiritual world of ancestors and consider them akin to deities. These rituals and practices of faith bind the community, enabling them to make many collective decisions and actions for natural resource management, thus sustaining livelihood systems and strengthening kinship across social and ecological communities.

The functional beauty of the four above-mentioned principles of kinship found among the Lepchas—reciprocity, solidarity, equilibrium, and collectivity—are not only available to Indigenous communities. They are available to all who want to cultivate them. Any community can utilize them in evolving the practices applicable to their situation. They can be integrated into modern-day community living and can reveal the linkages of the abiotic and biotic—linking people with one another, with plants, with soils, with mountains, or with other spiritually, culturally, and ecologically significant features that define and hold together a human and more-than-human community as kin.

Landscape Connection and Biocultural Heritage

One way modern societies have sought to support and learn from Indigenous kinship connection has been through international biocultural heritage programs. According to the website developed by the International Institute for Environment and Development (www.bioculturalheritage.org), biocultural heritage is "held collectively, sustains local economies and is transmitted from one generation to the next. It includes thousands of traditional crop and livestock varieties, medicinal plants, wild foods and wild crop relatives. These precious resources have been conserved, domesticated and improved by communities over generations—and sometimes millennia." Significant, indeed, but for those who are living in cities far away from production landscapes, as well as those who are aware that many of these traditional communities themselves are in transition to so-called modernity, this approach may sound romantic. However, there is considerable value in the kinship principles of interdependence and interconnectedness that Indigenous wisdom provides and that can be recaptured through food-as-wellness movements.

Awareness of Abiotic and Biotic Entanglement in the Age of Holistic Development

Kinship happens, kinning occurs, when we actively engage in entangling biotic and abiotic systems. It is well-recognized that Indigenous science in traditional cultures made significant strides in understanding how the abiotic and biotic influence each other. But how do we prevent this knowledge and attendant resource system from sliding away? The tie between physical and intellectual resources is often so critical that it is hard to keep one intact without the other—such as the biological diversity of food crops, or the cultural diversity of recipes and native plant-based medicines—and they cannot be maintained for future generations without any active use of these resources. Merging and rekinning systems previously reduced to abiotic (like water or soil) components within biotic systems such as plants requires us to merge the social and the ecological—the soil with the plant with the taste and sounds and smells of a family kitchen laughing over an ancestral recipe.

But how do more of us fit these into our modern lives? We must find ways to connect the next generation of modern leaders with the four kinship principles of these Indigenous food practices. We must see these principles as a doorway through the false dichotomy of abiotic and biotic food systems (like my day looking at roots in a hydroponic tank). One of the pathways could be helping young people worldwide to live with, partner with, and learn from these communities in their homes.

My Foundation for the Contemplation of Nature (foundnature.org), based in the Himalayas of northern India, developed a course with Western Colorado University that introduces young people to these principles. The course, "Mountain Resilience," opens food-based portals into soil-plant, social-ecological, and abiotic-biotic kinship. It starts simple, with American students living and helping on an Indian family farm for a month, which includes the act of preparing and (yes!) enjoying a meal. The simple act of preparing soil, growing food, milking a cow, gathering fuel wood, cooking a meal, and sharing stories around a feast (a celebration of cross-cultural kinship) opens up other doors of kinship that go into making a social-ecological community resilient—health care, holistic well-being, architecture, town planning, and even maintaining caring human relations among people, soil, flora, fauna, air, water, and the atmosphere shared in any given landscape. Many students realize—just from one meal—that without the four kinship principles of reciprocity, solidarity, equilibrium, and collectivity, they would have gone hungry. And without this four-layered kinship necessary to fuel their bodies, they would have suffered both physically and spiritually from reducing the land and its people (biotic beings) to mere resources (abiotic beings), thereby reducing themselves to mere consumers of a falsely severed planet.

Conclusion

I know this not only because I have seen dozens of students transformed by moments like this. I, too, remember seeing food in a new way, through the eyes of kinship. I was working with rice farmers in the Apatani landscape of Arunachal Pradesh in India. My university education taught me that any nonrice plant in this particular food economy was uneconomical and thus should be killed with herbicides (along with all the insect "pests"). Then, one day, I entered into the community and met a woman weeding in a rice field.[3] There were many types of nonrice plants but she was removing only a few. I was taught that those nonrice plants were useless, to be either killed or readily uprooted and thrown away like rocks (which I have since learned are themselves living and vital to our kinship within living systems). I asked her about them. She smiled, honored to share her wisdom, and said:

> These ones that I am removing are edible green leaves for dinner today, those ones that I am leaving behind have many uses. For example, some of them are liked by the insects and worms, that in turn feed many fish inhabiting the waters of this patty field. We don't have to feed the fish. The insects also attract many birds, which also feed on insects that damage the rice plants. The droppings of the birds are good for the thriving life of microorganisms that help with intake of nutrients through a healthy root system. And these rather taller ones that I am leaving behind have much deeper root systems than the rice and help improve the subsoil habitat. We turn them over after harvesting the rice and so on.

Although she did not use the terms I was used to hearing from fellow scientists, the intricacies of her wisdom left me in awe. Moreover, it's not just these smaller, nonrice plants that inhabit a paddy field. Naga farmers in the northeastern corner of India, for example, have perfected alder-based agroforestry systems in which trees that are over one hundred years old are present, kinning with rice plants that nourish them like grandparents.[4] The alder roots fix vital nitrogen from the air.

The way those plants—living together with rice—awaken soil health, plant diversity, habitat, and a healthy and beautiful human homeland has stayed with me. It has showed me that true kinship, true kinning, is the opposite of those reductive moments I experienced in the plant physiology lab in 1988. It showed me that true kinship lies in our organically experiencing and participating in nature's complexity as a part of the ecosystem, not in standing outside with instruments of measurement that can explain what we hypothesize only through the lens of efficiency.

NOTES

1. A. Rastogi, N. Gurung, and S. Sogani, "Smallholder Innovation for Resilience (SIFOR)," *IIED: London* (2014), https://biocultural.iied.org/smallholder-innovation-resilience-sifor.

2. Further information about these principles is available at the Biocultural Heritage website, https://www.bioculturalheritage.org.

3. A. Rastogi and R. Pant, "Socio-Economic Context: Case Study of Eastern Himalaya," in *Changing Perspective of Biodiversity Status in the Himalaya*, ed. G. S. Gujral and V. Sharma (New Delhi: British Council Division, 1996).

4. A. Rastogi, "An Inquiry into the Relationship between Agricultural Research, Development, and the Issues Facing Agricultural Biodiversity Management in the Hindu Kush-Himalayan Region," in *Managing Agrobiodiversity*, ed. Tej Partap and B. Sthapit (Kathmandu: International Centre for Integrated Mountain Development; Rome: International Bureau of Plant Genetic Resources, 1998).

THE COAL REMEMBERS
Trebbe Johnson

This creek has some kind of gumption. That's my reaction as I consider Sterry Creek, rippling along through the woodland between banks of refuse and neglect. Near its eastern bank, a bare, black hill of coal waste hunches its shoulders against the blue October sky. To the west, the woods along the pitted and stony track I'm walking are scattered with more contemporary trash: a weather-ravaged mattress; cans and brushes left over from a paint job, laid out with anomalous tidiness on a blue plastic tarp; shiny clots of fused stuff; lumber, tires, and more tires. And yet on flows the creek, as if it had somewhere to get to and were so intent on its journey that it bothered little about the detritus it must pass along the way. Narrow and clear, it lifts and shuffles cast-offs no more noxious than autumn leaves.

The big heap gleaming like an obsidian mountain range in the late afternoon sun is called a culm bank. It's composed of shale, sandstone, and other unsalable tailings that were removed from the ore during the coal boom in the 1920s, 1930s, 1940s, and 1950s—and it's one of hundreds left throughout anthracite coal country here in northeastern Pennsylvania. Although local families regard the culm banks as testaments to the hard work that their parents and grandparents from Wales, Poland, Italy, Ireland, and other countries came here to live and die in service of, the implications of them, if not the mounds themselves, are monstrous when you actually pause before them to gaze and consider. For those of us who live in this region, that's easy to avoid doing, because they are almost as much a part of the landscape as their daintier cousins, the green hills known as the Endless Mountains that start rolling north just a few miles from here. You get so used to coal waste that it ceases to be shocking, as some new ecological outrage would be, something like a city block blown out by a bomb, or the charred skeletons of houses and trees ravaged by wildfire, or the hole that opened up in 1903 in the town of Olyphant, at whose southern end this coal patch lies. When a ceiling of the mine collapsed, it swallowed the West End Hotel, pulled the walls of several neighboring buildings below the ground, and snapped the water main.

Over the decades, thirty thousand people died working in Pennsylvania's anthracite mines, and yet the mining went on, and the culm banks peaked higher and higher. *How many men, boys, mules, and vehicles,* I wonder, *must have labored up those slopes on the other side of the creek to erect such a store of blatant undesirability?* The women of the mining families would have picked over it in search of overlooked lumps of coal that they could carry back to heat their small and insubstantial company homes. The mound is a wound on the current landscape and a scar of the past. And yet the features that frame it include, right before me, a scrim of thistles, goldenrod, milkweed, and tall grasses, all in their midautumn fuzziness, and on the far side of the coal, the gold, crimson, and copper chenille of trees at their picturesque peak. Like the creek, they prevent me from assuming a stance of unmitigated sorrow or indignation about this place. It's not pretty. And then again, it kind of is.

Behind me, the traffic on Route 6 murmurs and coughs while crickets offer a friendly, prefrost greeting in the scrub. Along both sides of the track, trees of a variety surprising for a waste place shimmer in the breeze, and leaves sashay through the air onto the ground: golden tulip and mitten-shaped sassafras, red maple, ochre birch, leathery oak and ash. Flycatchers flit among the sunny leaves of poplars, commenting briefly to one another. Purple wild asters, yellow partridge pea with its showy red stalk, and the miniature daisies of fleabane are still blooming in patches of sunlight.

Abruptly the soil under my boots shifts from powdery and dough colored to gritty black, which means I've just stepped onto a vein of coal. Olyphant is near the tip of Pennsylvania's northernmost anthracite coalfield, which curves out from Carbondale in a long, thin shape, roughly resembling a pea pod, to taper off again at the town of Shickshinny about fifty miles southwest. I'm not walking on solid ground. Underground stretch labyrinthine passageways constructed in a pattern known as "room and pillar." The "rooms" were spaces cut into the rock, and the "pillars" were those parts of the substratum left untouched, so as to keep the whole architecture from collapsing. The miners worked these tunnels ten hours a day, six days a week. When a man died, his body would be dumped in front of the small house he rented from the coal company. Everyone in the family knew that if they didn't find a replacement for him within a few days, they'd be kicked out of their home. Little boys as young as seven worked as "breakers" for pennies a week, bent over those rocks that never ceased to pile up before them as they sorted the lumps into different sizes and picked out the waste that would add to the culm.

When the steep path reaches the far end of the culm bank on the other side of the creek, and the terrain to my left opens up into woods, a sense of relief floods me, as if both the land and I were released from a heavy burden. A large puddle lies to the side of the path, and an animal trail bypasses it and curls into the woods. Glossy black against its base of coal and tessellated with scarlet, gold, and bronze leaves, the surface of the puddle looks like fine Japanese lacquerware. A large emerald dragonfly darts purposefully among several invisible midair ports of call. One of the rewards of spending time with waste places is the startlement of beauty they hold in reserve. They take what has happened to them and deal with it. They are pushy and creative, and they muddle through, working with what they've got without expecting any favors. There is much to discover, both in the land and in myself, whenever I step into the mystery of a place that's been through a lot of trials, yet I almost always postpone the journey. On some level, I know that if I allow myself to truly experience what is before me, my worldview will be rattled in ways I can't anticipate. I will, at the very least, be forced to experience something, perhaps uncomfortable, in the tender inner frontier where "I" meets "other." I risk feeling sad, shaken, and mad. Once my body is on damaged ground, my mind will not be able to insist quite so vehemently that I know what needs to happen there. Therefore, my mission today is just to absorb what is. I must greet and engage with this place in its integrity, as I do whenever I settle in to a shy first encounter with what the philosopher Emmanuel Levinas called, in *Otherwise Than Being*, the "vulnerability" of the face of the other.

The track diverges a short way off to the right and comes to a sudden end before a tall chain-link fence. Behind it, the added prohibition of a camera mounted on a tall pole warns me that trying to proceed will not be tolerated. I smell smoke. Or I think I smell smoke. Lifting my head like a deer, I sniff and sniff again, but it could be that the scent I'm gleaning is only decaying leaves, or maybe just imagination. Then again, it really could be smoke, because although the vista behind the fence is an innocuous stand of young poplars shimmying in the breeze, I know that this fence marks the boundaries of an underground coal fire. I'm tempted to follow the fence a little farther, but the sun is already glowing through the trees in the western woods, so I return to the main track and turn around. Before getting in the car, I turn and make a deep bow to the whole place.

The only furniture in this narrow, windowless room of the Bureau of Abandoned Mine Reclamation in Wilkes-Barre is a long span of tables running down the middle, with chairs on either side. Affixed to the wall behind the chairs are waist-high racks from which hang large folios depicting the interiors of the region's coal mines. The mining engineer Dan Werner has selected one, Folio 10X, and laid it out on the table, but before he opens it, we bend over a smaller portrait of Olyphant's mines, an aerial photograph. Several patches of the land below the highway, dark woodland green interrupted by two gray humps of culm piles, are marked with yellow outlines. Each one, Dan explains, is known as a "PA," a problem area. The problems include subsidence, dangerous high walls, leftover refuse, clogged stream channels, old slopes, and the biggest problem of all, the underground mine fire.

It started in 2004, when somebody decided that a culm bank would be a great place to set an old car on fire. The fire did what it could with steel and rubber, then moved on to embrace a more compatible partner. Down into the coal mine it slipped and then outward toward the town of Dunmore to the west. Coal burns slowly, at the rate of about one foot per day, but even when it's slow, contained, and invisible, a culm fire is toxic, for it emits greenhouse gases, as well as carbon monoxide, hydrogen sulfide, and various trace elements. In an effort to contain the damage, the Office of Surface Mining dug a U-shaped trench 150 feet deep and 2,800 feet long and lined it with clay. That prevented the fire from spreading horizontally, but it's still smoldering throughout an estimated seven acres at temperatures ranging from fifty degrees Fahrenheit to more than four hundred degrees.

Northeastern Pennsylvania's mining industry ceased almost completely after 1959, when miners, under orders, dug too close to the banks of the Susquehanna River. The river breached the mine and the tunnels flooded, killing twelve people and injuring dozens more. Decades later, many mines farther south are still under water. Higher elevations in the north have kept Olyphant's mine dry but more susceptible to fire. The coal veins are also thinner here, as I see when Dan peels back the cover of Folio 10X. On each large page is a floor plan of one level of a particular section of the mine, each room and pillar meticulously drawn. Olyphant's folio contains only three pages. The folios of mines farther south, Dan explains, are eight or nine pages, each one representing a level deeper than the one above it. Around the town of Shamokin, the miners would have descended more than five hundred feet into the Earth each day to do their jobs.

These days, the only thing that's moving through those rooms and breaking them up is fire. The plan, Dan says, is to dig out the uncombusted coal and extinguish the fire, then fill in the trenches and reclaim the land and about seventy-five acres surrounding it, so it can be converted, perhaps to an industrial park. When that work might begin is unknown. Meanwhile, although mining in this region ended decades ago, the coal itself does not stand idle. Abandoned, it is yet active. Like Sterry Creek, like poplar trees sprouting in a black bed of carbon and a dragonfly busy with the air over a puddle, the coal goes on, doing what it does well. It burns.

A few days after my visit to the Bureau of Abandoned Mine Reclamation, I return to the coal patch. This time I take a different path, less traveled and surrounded by woods. It, too, is littered with trash. A paper plate nailed to an old wooden door that's been propped against a tree is riddled with bullet holes. A child's pink plastic pedal car lies among a dozen plastic grocery store bags, their handles knotted to keep the contents inside. As I watch, leaves drift down from the trees to slip and slide over the sides of the bags and pad the molded pink interior of the little car.

Waste attracts more waste. When a place is seen as useless, unwanted, uncared for, it loses value not just once but increasingly over time. It becomes a pariah of the landscape. Before long it is no longer a place in need of attention but a repository for other things that have passed from usefulness to junk. Once so condemned, it invites deliberate acts of aggression and disrespect, such as target practice, dumping trash, and setting an old car on fire. The place has become good for nothing—nothing except expressions of contempt. Knowing this sad evolution keeps me coming to places like this coal patch: I feel sorry for them. They have given everything that voracious humans demanded of them and now, spent, they are despised. Yet they are like friends who have fallen ill; although I can't heal them, I can attend to them, praise them, give them little gifts.

Higher, the path ends at another section of that forbidding chain-link fence. Animals have made a track around the foot or so of roughly horizontal land around the perimeter, as if they were pilgrims circumambulating a sacred mountain. Although the woods to my right are thick and the slope steep, I can amble along in a counterclockwise direction with no problem at all until I come to a large maple tree that has crashed down and completely blocked passage. Its upper branches have bent the top of the fence, and briefly my imagination plays with the possibility of climbing diagonally into the fire area. Instead I bushwhack through the woods to the place where the overturned tree forms a wall of mud, stone, and root parallel to the metal fence. A scattering of boulders on this rise of hill makes a good place to sit, so I do. I'm vaguely curious to see the fire up close. Mainly, though, I'm thinking about how alive this place is. I'm reminded again of Levinas and his writing about the open, needy face of the other. When I let myself be drawn into the vulnerability that's communicated beneath first impressions of the one before me, I soften. My defenses slip. It dawns on me that, wherever I look, I will see a life that is etched with some hurt, even if the scars aren't always visible. It's true of places no less than people. In that suffering and survival, I see beauty and discover something like love. Before I leave, I weave dried flowers, leaves, and a blue jay's feather into the fence as a parting gift.

My reflection that the trail around the fence was like a path of circumambulation around a sacred place gives me an idea I can't shake off. "Circumambulate the fence!" I scrawl on a Post-it note and stick on my computer. By the time I return, it's early November, and the sky over Olyphant is temperamental, reeling from overcast to bright sun and occasionally releasing flurries of snow. The puddles I sidled around on my previous walks are covered with a thin layer of ice. Except for the oak and a few enthusiastically twirling poplars, the trees are bare. Under the windy gray sky, the culm bank broods ominously.

I will say this about that walk: there is surprising beauty and gargantuan scarring, the land is busy the entire time, and my mood gets whipped around as wildly by the land as the trees are by the wind. Just seconds after I begin my walk around the fence, clockwise this time, I push through a thicket of spindly cherry trees and spot a large silver Christmas-tree orb glistening in a sudden shower of sun. How did a Christmas ornament come to be here? By wind? Dumped? How is it possible that it is so little battered after all it must have been through to get from somebody's holiday tree to the coal patch? Picking it up and nestling it in my backpack, I feel gleeful and triumphant, as if my good intentions for making this odd solo pilgrimage have been noted and applauded. Pride shifts into aesthetic appreciation as the hill levels off into a flat area, where poplars with white trunks and buttercup-yellow leaves sprout from the ebony coal floor sparkling on both sides of the fence. Deducing at first that coal is no deterrent to growth, I begin to grasp that all these trees are young and spindly, a clue that their ambition does not match their ability to mature in such circumstances.

As I peer at the vista through the fence, four ravens glide overhead, playing with the wind and one another. One veers over to investigate me. The birds drift onward when I make the turn to walk the far southern side of the fence, and at that moment the sun disappears behind a cloud. Thickets of blackberries now knot my path. Instantly I feel abandoned and edgy. I start worrying about people with guns and all-terrain vehicles and other hobbies of messy destruction. I chastise myself for forgetting my orange vest when it's now two weeks into hunting season. As I press on, ducking through the thorns with eyes closed and head down, my nervous musings attach to a friend who's just been diagnosed with ovarian cancer.

On the long western side of the fence, the landscape inside the containment changes abruptly. Just a few feet away from me tower mounds of coal waste, thirty or forty feet high. Deep pits run among them. There is no scent of burning, no sign of smoke, but knowing that this is the place where fire is slowly devouring the substrata is like standing at the site of a recent highway accident or act of violence: the terrible reality of the unseen coats the place, blotting out the apparent ordinariness of what's before the eyes. The fire in this colliery is one of an estimated forty burning in Pennsylvania's abandoned coal mines, both here in the east in anthracite country and in the western bituminous mines around Pittsburgh. Like the fuel rods of nuclear power plants, like garbage moldering in a landfill, like carbon drifting into the skies and sticking there, the life of a thing does not easily vanish from the Earth just because humans are done with it. It lingers, it seeps, it clogs. It keeps finding ways to be part of the Earth.

The lines of Yeats's poem "The Circus Animals' Desertion," which have guided me for decades, swim into my thoughts: "Now that my ladder's gone / I must lie down where all the ladders start / In the foul rag and bone shop of the heart." In other words, down at the bottom of as-bad-as-it-gets crouches the possibility of renewal. Compelled suddenly to make that paradox manifest, I take Yeats literally and lie down on the ground beside the fence. I'm half-hoping for another Christmas ornament moment—for the fire to reveal itself and radiate some gentle warmth over my back and legs, or at least for another flyover from ravens. It doesn't happen, of course. It never happens when you demand it. The ground is cold, the snow is brushing across my face, and the stones under the leaves make a punishing bed. Nevertheless, the very act of positioning myself here, supine on a place that's endured and still endures so many trials, shifts my relationship with it yet again. To lie down voluntarily on a place is to anticipate relief and solace. It's to take yourself off those resolute feet that hold you up and propel you all day long and give yourself over to a more extensive gravity. I succumb to rest. Beneath me stretch loose stone, cold grasses, and a patchwork of rooms and pillars where many, many people did hard labor. If any of them ever lay down on the coal, they probably didn't do so by choice.

The coal waste that lies scattered on the path and piled in the culm banks in this place is the Earth turned inside out, like the pocket of a great jacket. Such places cause discomfort. They unabashedly bare, for any who cares to look, all that is foul, ugly, and spent, and maybe contagious as well. They don't keep their unpretty bits hidden, as one is taught to do in polite company. Yet waste places like this are teachers. They coach me in how to persevere, no matter how bad it gets. They remind me that, if I'm a little bit patient, I'll quite possibly be granted a gift I can't possibly have earned. It could be a Christmas ornament, a dragonfly, or a moment of cold discomfort as I lie on the stony ground and picture the presence of those who worked below. Most important, a waste place reveals that the memory of the Earth does not discriminate between beauty and ugliness, value and trash. This abandoned landscape remembers how to seed, flow, bloom, and chirp. The tunnels remember the many lives that labored and even laughed in them. The coal remembers how to smolder.

GETTING IN ON THE MAKING
Maya Ward

He who does not trust enough will not be trusted.
—LAO TZU

Weaving is an old way of knowing. Losing our fur was quite an incentive for getting in on the making: to find a way to dress our delicate skin, to shelter this sensitivity. The weave is a knowing like life: a pattern, not yet conscious, emerges in the creative act. Weaving in many cultures is a sacred art, a type of magic, a spidery kind of skill.

We can all see, now, the holes, perhaps irreparable, in the web of life. In this age of ecological unraveling, the subtle, gentle magic of the weaver seems too humble a thing to be of help. And yet without it, I fear we will become utterly frayed.

Some say this time, exactly as it is, is an initiation into something entirely new. Can we trust enough to be entrusted with the true, strange, terrible way things are? I think every person is called to stitch together their samples, to quilt themselves into this new home. I doubt if there is a soul alive for whom this will be easy work. The needle must be carved from one's own bone. The thread will be nothing less than our sinew.

We care for ourselves through tending our connections. Our love for this world, our kindredness with it all, and the actions that arise from love—these must weave a vessel that could nest a new culture. Through everyday acts of attentiveness, from aligning with the other as kin, change will come. Practices of kinship involve a conscious restorying of our irrevocable entanglements. All things are born from this system of earth and sun, a system entangled among endless stars, the dying of which gave birth to the elements of all our bodies. The root of the word *kin* means "to give birth." Kin is our intimacy with our things, our big and colorful family emerging as and from the eros force: we are this, noun and verb. It's a wild and sexy thing we've arisen from. All things are woven into it; threaded, knotted, bound. And within the weave we dance.

The Dance

The cloud shadows are racing up Tugwell. It's an inelegant name for such a lovely mountain, smooth with forest green, patterned purple where the shadows of the cumulus ripple over and away. Not much more than a century ago, a Mr. Tugwell came this way. Seven mountains ring this place, old beings, older than the oldest names. We call them, if we call them at all, Donna Buang and Little Joe, Mount Bride and Mount Groom, Victoria, Tugwell, and Story Hill. Only Donna Buang is known by her old name, her Woiwurrung name; Donna Buang, "the body of the mountain." I sense a convoluted tale, dense and contested, woven here between these beings.

Mount Victoria, the rise on the side of the range crowned by Donna Buang, is named for the monarch who oversaw the theft of this place. This name seeks to claim old earth, yet for those who live here, Donna Buang is our mother mountain, lush, deep and dark. Her shadow side is what we see and rises high above the rest. Little Joe, bright and steep, silvery and dry, is her opposite, the yang to her yin. Bride and Groom stand at the end of the village and lend their ritual: Donna and Little Joe, paramours of long ago, make Story, the small hill cradled between them, the hill around which our village weaves.

That leaves Tugwell. Ah, yes, suggestive. Let's not be prudish. Lust, unfulfilled? Desperate or sexy self-pleasuring? What does it mean to imagine a lonely mountain?

But then there is time, and change, and the volcanic hills worn down, worn down. The yin, the moist sides, and the yang, the dry harsh slopes—they've gazed quietly at each other for a long time. Yin and yang face each other, and the cooling river runs between. The Chinese first found this pair of words to describe the opposing qualities of shaded and sunned mountain sides, and from that early seeing went on to observe the pattern everywhere, of how the earth's undulations shaped life, shaped mind.

These seven hills now have a small village wedged between them, but the hills pay us little mind; they have been telling themselves the old stories for eons. They sit peacefully and share tales of the old days, of volcano lovers, eternities of making and shaping.

Don't believe my story. Let these mountains be. Let them be lustful and fecund, quiet and calm, everything and nothing. Don't believe it, but let the length of their lives teaches what is needed. Borrow their breadth and depth, be the idea of mountain for a time, try on tree, or animal, then let go of them all, be the infused and the empty space.

At yoga class, downstairs in the old Mechanics' Hall, we stand in *tadasana*, mountain pose. We move on into tree. We flow from pose to pose, becoming them so briefly. I can almost hear Little Joe snorting his disdain.

Or was that snorting the Sambar, the deer that are wild all around here, that thrive in these hills, brought over from India?

Right now, the stags are fighting, antlers clashing, every year growing them and every year losing them, and, yes, there is a loser, for this is a battle for sex. The stories of the first people, the old cautionary tales, they often have stories about sex, there among the mix. The enormous creative flame in every single atom of the universe, wanting to make itself, make itself, to be alive.

I have a pair of antlers on my shelf. I found one on my walk around the seven hills, following the creek down off Donna Buang.

We were upstairs at the Mechanics' Hall, the ash floors shining with new polish. The river just behind, the hills all around. It was what we did every Friday night: some girlfriends and I hired the hall every week and facilitated an open space for the community to come and dance. Some of us explored the dance form of contact improvisation, the play of weight, an exhilarating entanglement of balance, shift, and fall. Dancing with my friend, I draped across him, soil over a stone mountain. Then comes the landslip—I fall from his back and tumble toward the floor, but I am caught and twirled; this is the dance. The one rule of the game is to always have contact with him, be it torso or toe, brow or limb. We start fast and wild; we play by the edge, precarious, unbalanced, a system together of trust. Momentum keeps us spinning, we spill and sweep through the room, across the sprung floor made of these mountains' trees. We splash and race, I pour across his slopes, flowing between the yin and yang planes of his body. I am water across his warm brown flank, the ridge of his spine against mine. Then his weight over me, I flow out from under him, around him, over him, then under again—crushed, released, raised, and swung. Cradling and cradled, pressed and flung and swept and eventually spent. All energy is gone. But we were mountains together, we were water. We danced the elements, becoming and unbecoming, hinging off each other, unhinged, rejoining, until slowing, deepening into home. Eros is sensed yet not claimed, it is released to the dance. We bow to each other, we leave, each to our own place in the valley.

The Work

The clouds are dense over Tugwell. They're loosing their water, the mountain is getting a pounding. That inelegant name, I've just found out, has a meaning. It was lost on someone like me, divorced from my ancestors' skills and knowledge.

Names shift and change over time; the so-new name of this mountain was given an old word, bent from the original over centuries of language shift. Mr. Tugwell's ancestors were almost certainly once named Tucker, and a tucker was one who, after the cloth was woven, softened the rough-spun wool into something better for the body. To tucker cloth is to felt it slightly, making all those separate strands of the weave into one skinlike thing, warm and resistant to water. Over time and in most places the work of the tucker became mechanized, but in remote areas of Scotland the manual process continued. There the tuckers, known as the fullers or the waulkers, were women. Around a large table, women would gather and waulk the cloth with their hands and arms. To make good cloth required consistent pressure, so to synchronize the work of many hands, they used the rhythm of song, chants of call and response. The soft slap of the wet cloth upon the wood made the beat, and they chanted back and forth across the boards. To make it even was the thing, so they turned the long strip of cloth, passing it from hand to hand, clockwise around the ring of women, so that all women handled the entire bolt. This was the way differences between hands were smoothed, for all hands waulked all the tweed. The chants they sang were old tales, or improvised yarns of their days, their loves, their lives, told line by line, each line sung back to the teller: call and response, call and response. They sung their lives, their kin, the land, the sea. The long hours grew smooth and patterned with chant, the beat held the task in a kind of dance.

This was the way of the work, sustained in some places until the 1950s. They filmed it then, and so there is a record of it, and I listened to the songs and watched the women at their work, the susurrus of the cloth as it turned sunwise under their hands, the thud of the cloth as together they beat it onto the boards, the hypnotic song, the cloth sung, the work *enchanted*. For the word enchanted means to be inside the song, and to be inside the song changes the brain, it makes brain and being one rhythmic thing, it is the way of many of the old rituals of work, patterning the long and necessary labor with something akin to bliss. And together, for together tasks such as this were done.

Mechanization stole away the place of these rituals. Without the work, the songs were lost, the rhythm for singing of stories was gone. This happened all over Europe, the places where my ancestors lived, long ago.

The tucker did to the cloth what the rain does to these hills, softening and smoothing them, in rhythms faded from memory.

Our language is old, and times have changed, so we don't know what is inside our words. I'd thought the name Tugwell somewhat ugly, slightly lewd, but now that I know the story of the name I see it linked to old ways, old ways of women working, and I wonder what songs have been sung through these hills, sung for thousands of years by women at their collecting, at their weaving, spiraled baskets and eel traps, their sewing together of possum-skin cloaks with needles of bone, with thread of sinew, the making of the things of their lives.

The Night Forest

At the end of a sweltering day, I watered my veggie patch, harvesting as I went: parsley and basil, lettuce, rocket, cucumber, and cherry tomatoes. I made a salad from my findings, gathered some blackberries for dessert, and roamed, restlessly, around my garden. Then my ex-lover called me. "Do you wanna go for a walk?"

"Yes!"

We headed up my road toward the near-full moon rising in the east. We walked the paths to the aqueduct and followed that way further east, deeper into the mountains, toward the headwaters. We walked for hours. We walked under the moon upon the mountain. In the unusual heat of the night, a thin dress, my shoes flimsy, I felt light, ungrounded, as if I could float up into the gum trees. Eventually the mountain's flank leveled a little, the gum forest lay open and undulant under the moonlight.

We sat there in the black and silver forest. I could feel the mountain, a presence in mind and memory and actuality, a bulk beside us, earth risen and us leaning into the side of her, sheltered into her flank. The pillars of the gums made an ancient architecture, moonlight-cool. A breeze passed through. From high above came the whispers and sighs of their swaying.

We sat within old and unknown stories. Old stories, ancient patterns. Same pattern, different beings. I thought about the Celtic gods of earth and place. The Cailleach, a creator spirit, the powerful woman who made the Scottish mountains, rivers, and lochs. Sometimes she is figured as a crone carrying stones to build cairns. The mountain we sat upon, Donna Buang, part of one long granite rise that continues to the next peak named Ben Cairn, *ben* being Gaelic for mountain, *cairn* for, I imagine, the huge boulders at its summit. Flowing off Ben Cairn is a tributary of the Yarra, the Don River, likely named after the river Don in Scotland or Yorkshire, that name derived from the Celtic mother goddess Danu. Danu also gave her name to the river that snakes through much of Europe— the Donau, or Danube. And the word *Danu* can be traced all the way back to Sanskrit, where it means "fluid."

When I think of those names, carrying as they do the cultural history of my human ancestors, I imagine how the colonists may have sought to experience the depth they knew in their home places by bringing their old sacred names to where they had come to live. Farming folk in the main, driven off their lands in the clearances or the upheavals of the Industrial Revolution, in shock at the strangeness of where they found themselves, seeing meaning, seeking safety, seeking home.

Many millions of years ago, Donna Buang was a volcano, part of a string of giant calderas so powerful that their eruptions impacted the path of life on Earth. Giant plumes of dust entered the air, and ash blanketed land and sea, suffocating existing life, yet providing the nutrients that would sustain future forms, all the way down to now, becoming the soils of my garden. Every day I eat from the body of her volcano days. My body is what she once was.

Within the night forest, I could feel how the erotic life of matter extends even to stone. Lava is stone in its liquid, generative form. As it is ground to dust, its rich mineralized being becomes incorporated into all. To feel the living land is to be animist by experience, by feeling the kindredness of what I am and what I am upon.

Donna Buang has weathered slowly to become this quiet green giant, restful in elder years, yet still fecund, unfinished, ongoing. I have imagined myself into the earth below the mountain. I have felt the heat of the bed of lava beneath. And while there are no volcanoes in this part of the world now, inevitably they will rise from far below the stone again, perhaps hundreds of millions of years into the future. Passion here will once again take lava form.

To imagine it as less is not enough. To truly see what we are among is allowing the richness of our feeling to land, to let it fall into and through what made it and what it is of. Within the night forest, every out-breath is an offering, every in-breath a gift. Invisibly we give and receive with the trees. Science might tell us the hows and whys, but we know the exchange to our bones; it is what love shows.

The erotic can be felt in our bodies' desire to remake itself, in the psyche's desire to heal itself, in consciousness's desire to know itself. All of these pull us into intimacy. Life has need on many levels. We live those needs as our own—obviously, naturally—they are ours to live. But not ours only. Such knowing is partial, and therefore inaccurate. It hoards the will of life to the human. Such smallness makes it harder for us: we think we're alone when patently we're not. It confines us to a human-only world. But by aligning to feeling, allowing it to lead us into intimacy, by connecting with the way things are kin to us, is to live another way.

Our generativity is of the same type as the earth's. It is of the earth. Our stories are a human telling of a general condition. All the animal of us, the best of us, our wild and feeling self, she dreams, he dreams, to live the earth's ways.

We stayed there for a long while among the forest. Below our feet the mycelia webbed together the brains of the trees into a pattern of thought extending many hundreds of miles to the north and the east, threading white and woven throughout the Great Divide. And in the above world, the day birds dreamed, the owls and foxes hunted, the sugar gliders, sailing between branches of the canopy, drank from the night flowers. The one who once broke my heart stood tall and pale and silent. He was as beautiful as moonlight. I stood beside him, watching and feeling. We stood without touching, brimming with desire.

The Unraveling

We wove to shelter our extraordinary sensitivity. We wove cloth and we wove narrative. We wove ever-greater patterns of protection; shelters, walls, nations, wars. To shed false skins seems an immense risk, yet there may be no other way. At its wildest, eros is the will to trust all things, to be kin with all things, even in this terrible time. It is its own strange truth. It is love and naked fury. To say yes to the fray, to let go of the woven, to be an act of unmaking.

From this we will be made.

NURTURING THOUGHTS
Tiokasin Ghosthorse

Language is the foundation of civilization.
It is the glue that holds a people together.
It is the first weapon drawn in a conflict.
—FROM THE FILM *ARRIVAL* (2016)

The winds toss the brown autumn leaves in the chilled morning air as I walk with a mentor, Birgil Kills Straight, of the Lakota Nation, along the barbed-wire barriers. We linger behind our group of international visitors who have come to Auschwitz-Birkenau, Poland, the largest concentration camp during World War II, to spend four days of silent remembering for those who perished during the "Final Solution" from 1940 to 1945.

It was here that I began to ponder, listen, and then question: Why would anyone seek a "Final Solution" to exterminate others? On this annual pilgrimage of remembrance, our group of Zen Peacemakers walks the perimeter of the barracks, ringed by electric fence. We stop at various points along the two-mile route, offering words and tears. Our group includes the children and grandchildren of those who were exterminated or those who survived the death camps. At the stopping points, I placed *canli* (tobacco) to express my thoughts—in the traditional manner of laying tobacco on the Earth—alongside a question: "What can be done to create a common knowledge and shared understanding among diverse humanity?"

I think humans struggle with the meaning of knowledge, often believing themselves to be its sole owners. Might we connect to another way of knowing, in which understanding is derived from watching how life nurtures living?

At Birkenau, we walked through sections of the camp that were oddly named: "Canada" and "Mexico." Canada contained the warehouses for all the things the Nazis took from the prisoners. Coats, food, kitchen implements, luggage, eyeglasses, false teeth, wigs, shoes—all the valuables that had any material worth, and everything else that could be salvaged. Europeans in those days considered Canada a place of wealth and prosperity. So the camp's authorities called this place of pillaged belongings Canada. The prisoners who got to work there had a privileged position of surviving by finding food and other things from what was taken from the luggage. Mexico was the area of the camp to the south of Canada where they would haze the prisoners after they were stripped of all valuables; it was called Mexico as a reference to lawlessness.

My mind abhors borders, walls, and barriers that are created to divide and conquer, whether on land or in language. I seek to understand the origins of perceptions that find only differences between us and promote indifference to the experiences of others. Such barriers were fully on display in this section of Birkenau.

The sun was high as we walked around the perimeter of the camp. Between "Mexico" and "Canada" our group gathered in a circle. The four Natives—a Lenape-Shawnee and three Lakota—came forward within the circle to offer a *canunpa*, or a pipe ceremony, for what we had heard throughout the morning. As Birgil filled the pipe, and I sang the four-directions song, I noticed the enormous flocks of birds, rabbits, and a small herd of deer in the distance that had negotiated their way between the cordoned-off barbed-wire fences into a no-man's-land where humans were not allowed to wander. These other animals had ignored the boundaries created by humans.

In my experience, when Native people gather in a circle for ceremony, oftentimes others naturally gather in this circle formation as well. I have wondered if it is a mark of respect for our ways, or if the people who gather feel some distant memory, within themselves, of the nature of the circle.

So, there I was, holding fundamental thoughts, fundamental questions, seeding relationally into the circle. I felt fully aware, seeing and feeling, noticing every facial movement and hand gesture among us. Then, like the lifting fog around us, there arose a knowing intuition, a medicine moving among us, as if the others understood a memory without having to mouth the words. In this circle, everyone was equal, balanced, quiet, respectful, and conscious. The birds stopped flying, and the deer and rabbits remained still, observing the participants.

I began to sing in the Lakota language and, in that sound, the people, the land, and the world seemed at once in unison—a kinship of all life. Within that circle, there was no feeling of need for anything else. The knowing of being nurtured became *realized* within the circle. Reverence for all life and all death encircled us. The power of the circle became the balance of relationship, which was both beyond and inclusive of the tragedy that took place on the land we were standing upon. I think that singing in a language of energetic memory brought all those present—human and nonhuman—together to honor those who perished, and also brought an integrity to the Earth that witnessed the memories. The past and future are all within the present, enfolded in layers of multidimensional reality.

This way of knowing and being is carried within the Lakota language, a language that verbalizes the energy of relationship to all, and then verbalizes the motion of that energy. For example, the word *wakan*, or "pure energy," is often used to acknowledge the presence of the mystery in everything. This presence of mystery might be understood as the foundation of our kinship. Our ancestors understood the effects of being *within* the metaphorical world of quantum physics, where there is no division between any dimensions of the world, where all is present and interconnected at once. Crazy Horse—Tȟašúŋke Witkó "His-Horse-Is-Enchanted"

(c. 1840–September 5, 1877)—a Mniconjou and Oglala Lakota, was quoted saying something that gets at this truth: "We live in the shadow of the real world."

No matter what language you speak, a hidden, nurturing world has already discovered you and is already at work with you, whether you know it or not. Humans have already been given the gifts, tools, and potential for this relational knowing and yet this source often goes unacknowledged, because we lack understanding of the sublime intelligence of Nature's nurturing, the nurturing that becomes the human body, that *is* the human body. Although we may speak of lack and yearning, I acknowledge that this already-present, hidden world is already listening to you and meeting every need with abundance. I also acknowledge the whole process of the sun—as a verb—as a giving, living being; sunning through the trees, rooting, leafing; creating consciousness; and finally producing the reciprocity of life-giving oxygen, as we return carbon dioxide to the trees, which, as a process, is a form of acknowledgment to the sun.

At Auschwitz, within our circle, there grew an awareness of consciousness known to some as respect for the seeing and the unseeing worlds. In that moment of remembrance from the past and the future, an entire history moved into a fulfilled circle of reality that we can choose to live with or not: the grief of human history. This "sense of fulfilling" was without dogma or religion—it was replete and relaxed—and in that circle, I recognized binary thoughts such as war and peace were impractical. I am reminded of something an eighty-six-year-old elder said: "When I am in prayer, in *wakan*, I am never hoping or promising, nor wishing something to come true. Only when I am in the moment of prayer acknowledging is the power present to move the impractical."

I looked into the eyes of those in the circle and could see men and women shedding tears while smiling. There was a sense of relief, a sense of grieving—a conscious common understanding of life-giving motion in that moment that struck me as innocent. And within that innocence was a fresh start, forgiveness, the knowing that all things are always in *wakan* (the mystery of and in everything), and most important, knowing that every moment never exists again. It is our responsibility to expand this realization, to acknowledge the consciousness of the world, not through a how-to instruction manual but by recognizing the living power of Earth and respecting all living beings. Mother Earth, as a being, is always listening to us.

Bullying Nature

What languages are the foundation of sustainable cultures without "development" or "progress"? Let me rephrase this. What languages are sustaining cultural relations with Earth?

Yes, there is a communication with Earth, and we have been detoured away from it by anthropocentric perceptions and a language that, for the most part, neglects the Being that sustains and nurtures all of life's forms. What rationale excuses us while we continue to extract the sources of life that remain in the Earth? Is there an exit strategy for anthropocentric languages? Will we realize that the same languages we use to try to awaken are the same languages that put us to sleep? Is there a language that has always remained in an awakened state?

I'm reminded of a story by Vi Hilbert (Upper Skagit) that has been passed on to encourage respect for the Earth. I heard this short story at a college graduation event in 1994. Vi noted that, in her language of Lushootseed, people cannot call one another a skunk or a dog. That would be disrespectful to the skunk or the dog. So often, we indirectly blame our lack of responsibility on nonhuman animals and detach ourselves from our direct responsibility as humans—individually or otherwise.

If we continue to use domination and dominating languages—enforced by bullying or policies of education, government, and science—and acquiesce to those same authoritative belief systems, then we will continue to feel detached as a species. We will continue to act as though we were separate and did not share relationships with all life—plants, trees, stones, fire, water, species of living forms with legs, and those with wings of the air, wings within the water, and those that crawl and creep within the earth.

A relational language is a language without nouns. It is a language that does not need to add suffixes to explain concepts and theory. It is not founded on subjugation; rather, it springs forth from kinship. A relational language never disconnects. It is always conscious, knowing respect for all forms of beings.

This sounds idealistic and implausible when our default language stems from domination. But think about a language that is only relational—a language that has evolved past the need for an alphabet, the need for nouns, the need for a beginning and an ending, the need for time, the need "to be" or "to become" someone other than you are now. A language that has evolved past the need for concepts of domination. Consider a relational language that recognizes the ever-moving *skan skan*, which, in the Lakota vernacular, is the continuum of motion behind the motion.

I wonder if Westerners know there are Earth or Indigenous languages that already understand the forces and principles involved in the living world, having evolved from thousands of years of experience and observation. Sooner or later, the perseverance that Native peoples have exercised over thousands of years—waiting for this time to arrive—will result in finally having their languages acknowledged and recognized as languages of sustainability and longevity. These languages are based on observing how animals and plants have adapted and maintained their interactions as *wamakaskan* (all living forms on the surface as sacred to life beneath ground level). Native peoples, like their languages, have also adapted, moving with the changing Earth, and, for example, planting seeds according to the land's capabilities rather than adjusting the land to fit our needs.

Original Intuition as Medicine

It was in Auschwitz-Berkinau, a time away from the vast plains of North America, where I came to understand relational values through the simple yet profoundly articulated First Nations language of the Lakota. It was a question I asked of Birgil: "Do we have a word or concept for domination?" He responded with a sound of quantum physics in motion—a simple "No. Domination does not work in a relational language that has evolved beyond war and peace unknown to the human race."

Further into our discussion, I understood the way "relational languages" are critical for learning the balance, rhythm, sound, and motion of Mother Earth. So, when Birgil and I conversed in Lakota, there was no "domination conversation" possible within a thought process based on the Lakota language. The separation between Lakota and the land, however, was transparent when we spoke in English. We described the energy, and the motion of the energy, that could move us through the separation we felt so as not to stagnate the *wakan* and to consciously apply mystery to everything.

I am reminded of when I was a boy sitting between my grandfather and grandmother, hearing them speak in the motion of old Lakota language, where there was a difference in how the female and the male referred to all life, to each other, in a relational manner. I would think on how they spoke to each other in a gentle, calm, and effective way, feeling an intuitive healing once I understood the deeper meaning. They would heal each other with the language. The energy, the quantum physics, the verbs that made everything come to life, felt never-ending.

I opened this essay with the epigraph "Language is the foundation of civilization. It is the glue that holds a people together. It is the first weapon drawn in a conflict." This assertion makes a clear point: language is not neutral. It can be a bonding agent, strengthening the ties between seemingly disparate things, and it can be a weapon, dividing things that are more alike than not. My experience at Auschwitz-Berkinau helped me understand how language is a force that can heal, speaking our relations into being, becoming a source of life, gathering together our kin. The Earth is listening. It is time for us to learn to speak what she is hearing.

KINSHIP WITH TREES AND CROWS
Alison Hawthorne Deming

To be rooted is perhaps the most important and least recognized need of the human soul.
—C. D. WRIGHT

When I think about what I miss most
what I wish to be in conversation with
at my summer home in eastern Canada
from which I'm exiled by the pandemic
and our government's failure
to protect our people unlike our kinder
northern neighbors who have done well
to shelter themselves and each other
what I miss most, calling to mind the land
I've known since I was a child
are three smart trees and a kindness of crows.
I refuse to call them a murder. I wonder
if they sense my absence? The crows
I mean. Should be the same population
summer after summer that speaks
in my dooryard, the caw-caw of their
intelligence familiar and inscrutable to me
yet familial. When I arrive I see five or six
gathered in the tamarack snag that I call
the Crow Tree. They love its bare
upper branches reaching as if choreographed
to gesture for the sky, one arm straight,

the other bowed so that together
they look like some kind of implement
a primitive pitchfork or a partial trident
or a giant's gaff hook. You get the picture.
I'll throw a caw or two or five
in their direction, trying to match the number
of times they iterate the syllable. That's about all
we can say to each other. They pretend to ignore me.
But I know they know me and that I am part of
the rhythm of their year as I arrive and plant seeds
and spread some extra for them to eat. That snag
is the first of my three smart trees. It refuses
to fall, dead now for a decade and its companions
long cleaned out with chain saw and splitter
after the infestation of bark beetles turned
a forest into a field of death. Our pandemic
is novel only to the human eye. Even the crows
suffered such a scourge when the avian flu
burned through them like wildfire. That's when
ornithologists learned that no crow dies alone
a companion always by the side of one dying.
The tamarack has a purpose as launch pad
for crows, as object of my admiration,
as place of reverie and prospect
for all of us in the small interspecies
family who dwell here. Then there's
the black spruce, a lone survivor
of predation's purge, a tree too broad
for my arms to encompass. During
the purge it lost its spire, standing
blunted and doomed for years
telegraphing crown decline which I
took to be a sign of its death.
But slow by slow the spruce trained

an upper limb to migrate from
horizontal to vertical. It seemed
as surprised as I was at this sign
of vitality. The makeshift spire
sprouted a mass of fresh cones
a gonadal extravagance or boast
that filled me with a strange erotic
joy. Finally the old cedar, rare on this
working island where fence posts
and shingles have been the plight
of cedars for two hundred years.
I imagine the farmer who once owned
this land prized the copse of cedars
rising up from swampy ground
at the perimeter of his stable corral.
Oh but that's history when I've been
on this land for more than sixty years
the farmer and stable gone decades
before that. But the cedar's hung out
for all the transitions, from my childhood
laughter to my old age astonishment
to still be alive and well and rooted here
even if seasonally transient. I like the way
the cedar is dying. It's been leaning
seaward year by year, now at forty-five
degrees from fallen. Its root-ball
is lifting and along its skyward
flank new branches have sprouted.
It's reaching for light but I think also
working to counter gravity's pull
preparing to become a nurse tree
when it finally loses its grip. The death
will be gentle though the force
of such a weight hitting the ground

will echo and make stones tremble.
There will be a whoosh when the still
green branches touch down as they ease
the last passage from a life seeking light
to one embracing dark loam of earth.

FOREST: A FESTIVAL OF FRIENDS
Sunil Chauhan

The mist was hanging low and embracing the cedars, oaks, and maples while the humming of the cicadas resonated through the woods. The moistness had made the bark of the oak softer than usual and its touch was gentle. The moss and lichen were thriving and the path was dotted with maiden's hair and dog violet. The occasional sound of the tit, bulbul, and woodpecker added to the symphony of the breeze that made the mist dance with the leaf—the cicadas, though, continued unabatedly. It was a perfect setting to visit some good, old friends in a space that can be described only as home. As I sat under an Oak, I heard it whisper to the Cedar, "I say, they need her, and yet they bleed her." The Cedar replied in a deep voice: "Out here in the woods, doubt turns to faith. This is home to us all; we dance in the spring and whisper in the fall." Insight is like silence singing to our souls. A deep healing begins when we find our friends in the woods, filling us with joy, love, and hope. Gratification is the gift of the forest; gratitude could be our offering back.

The mountain likes its solitude, but it loves to meet the clouds. It longs to hear the stories of the sea and the sky. The cloud is the raconteur that wanders through the wilderness, journeying to share the essence of our oneness. The cloud knows that the mountain is a good listener, and like all good storytellers will tell you, a good listener makes a good story into a great one. The story goes like this: "The mountain has a memory of being the sea and the sea holds a mountain song that resonates from deep within. They remember being one and now they find their union in the movement of the rivers to the seas and the clouds to the mountains. While the cloud brings stories, the rivers carry all that the mountain has gathered from the clouds and from the valleys and plains that it traverses. The river carries to the sea the patience of the mountains in a flow that eventually brings home the realization that perhaps stillness lays in movement. This is a deep friendship and one that connects the dots to the whole."

Everything is living in relation to something. Life is an aggregate of the sum total of our relations and connections. We move through a space-time continuum in which every moment is interconnected to a chain of events. Every chain is part of another chain that circles, cycles, and spirals through different evolutionary pathways—all connected through a common source.

Are We the Seed of Evolution or Its Fruit?

To take a brief look at the journey of our evolution, we need to go back millions of years to our common source, a single-celled organism. Technically speaking, we all have a common ancestor and everything in nature is related. As Gregory Ryan notes: "No reliable observation has ever been found to contradict the general notion of common descent. It should come as no surprise, then, that the scientific community at large has accepted evolutionary descent as a historical reality since Darwin's time and considers it among the most reliably established and fundamentally important facts in all of science."[1]

We are all kin. Being in nature is like being home with an extended family. Like all families, there are moments of togetherness and moments of disagreements, moments of peace and moments of discord, but most of all are moments of joy, care, and love. Humans have evolved in the lap of nature along with all other species, which (in all manner of scientific technicality) we can call our kin. "Biophilia, if it exists, and I believe it exists," writes E. O. Wilson, "is the innately emotional affiliation of human beings to other living organisms."[2] Our genetic coding is mapped in nature, and hence we feel most at ease when we are amid nature. The inherent love for nature (biophilia) makes us want to deeply connect with other forms of life. Kinship is a natural tendency in us. In the words of Shakespeare, "one touch of nature makes the whole world kin."

Love and kinship are wonderful notions, in the light of which peace and harmony flourish. However, as was already stated, humans are a complex mix of paradoxical values. While we live to love, we also love to differ and disagree. Our values are as diverse as our species. While some lean toward frugality, others relish extravagance; some prefer humility, others thrive with arrogance; for some, equality is indispensable to human well-being, while, to others, power and control are the essence of growth. With such contrasting values, our lives are bound to bring challenges that can be overwhelming. Herein also lies the scope for our collective growth. One of the big challenges for us is to create space for acceptance and assimilation. This comes through a set of values that become the fountainhead for our collective endeavors. We know how much effort is required to sustain relationships, and this is what we need to learn as a collective: building meaningful relationships and sustaining them.

The Healing Forest initiative, which I helped found, is one such attempt to reconnect people with nature and to discover the healing powers of nature. Our intention is to reestablish the kinship between humans and nature in order to live more harmoniously for a healthier society and planet. We wish to reconnect people with nature through a range of nature-based activities, games, meditations, and walks. The philosophy of the walks is fairly simple: *think less and feel more*. The walks are a means to experience moments of deep calm, creativity, and clarity. Our approach is to slow ourselves, to know ourselves, and to grow ourselves. The idea is to learn simple life lessons from nature that can bring us closer to the core values that make our lives more harmonious and peaceful.

Our goal is simple: helping people heal, helping nature heal.

Forests are known to have great healing properties for our body, mind, and spirit. Some of the researched benefits that accrue from a proper nature connection are as follows:

- Improved focus and attention span
- Heightened creativity and self-confidence
- Improved mood and energy levels
- Increased empathy and emotional intelligence
- Reduced stress and anxiety
- Enhanced sleep and mental health
- Tool to fight depression
- Increased awareness and willpower to overcome addiction
- Accelerated recovery from surgery, illness, or trauma
- Reduced high blood pressure, lung, and heart illnesses
- Boosts to vitality and immunity

The founding principles of the healing forest walks are silence and being slow. Once you slow your pace and become silent, you are able to immerse yourself deeply in nature. Through this immersion, a range of benefits follows for your physical, emotional, mental, and social health. We reach a state of balance and harmony through a sustained process in which the mind moves from attention to awareness through observation and engagement. With greater awareness there comes greater clarity, and answers then begin to follow. The final stage of growth is meaningful action, where applying this understanding in improving our relationship with nature and society becomes integral to our collective well-being.

One of the core mechanisms of nature healing and forest walks is engaging the senses with the surroundings. This enables us to get into a meditative space, where insights from nature assist us in learning the lessons that are present all around us. I remember once imagining myself as a leaf on a maple tree; during a storm, I hung dearly and intently to the tree, which I considered my home and source. As the winds intensified, I could barely cling anymore. Suddenly, I snapped from the tree and began to fall, and as I was falling, I realized that I was actually floating. It was a surreal feeling—and then I touched the ground, and this touch brought peace. I quietly lay at rest with my fellow friends and family. Over time, I realized I was slowly transforming into the earth, the very foundation of my home, and this realization brought closure to the feeling of separation from which I suffered.

Nature is accessible in some form or the other to all of us—a lush forest, a lake, a river, or even a garden. Being in a forest or amid pristine nature is not a precondition for a deep connection. What is important is the intention and the desire to connect. A city park or a water body or even a flowerpot at home is all it takes to create this connection. The eventual goal is not just to seek a forest around you but to find a forest within you.

Let us a take a walk. A walk that connects us to nature. A healing walk. Walk slowly and silently. Let us begin by observing all that is around us and make a note of what it is that connects most with us. All we need to do is to bring our attention to the present moment. For example, as you walk, hold a pebble in your hand and focus on sounds. Stay with the sound that you hear, and every time you lose the sound, shift the pebble from one hand to the other. Try to see if you can stick with the sounds that you find. If you naturally lose the sound, so be it. However, if your mind begins to wander, shift the pebble. After fifteen minutes of this kind of mindful soundwalk, see how many times you shifted the pebble.

It is natural for the mind to wander; however, when we slow down and become silent, we allow our attention to come to the present moment. Take a moment to connect with all that is around you and, through this process, all that is within you.

Staying silent and slow lets us observe all things very closely. Bringing focus to the sense of sight, let us look at things as if we are looking at them for the first time. Take your time, observe, engage, and absorb. Once you have looked at another entity closely, before turning away, look at it again as if you are looking at it for the last time. The ability to bring our attention to the present moment and see things—and not merely look at them—makes us more conscious of the space we are inhabiting and our relationship to it. Nature simplifies the process of being attentive and aware. Our ability to observe deeply and engage meaningfully with nature helps us to become more aware of the kinship we share with the natural world. We can then begin to explore the myriad aspects of our kinship with nature and see it as a mother, friend, teacher, fellow student, wanderer, healer, seeker, lover, caregiver, or perhaps the divine. Many such doors of kinship open themselves to a world of limitless connections. Being aware of this kinship is the first step in creating harmony with nature and making friends in nature. As we deepen this awareness, we begin to feel more gratitude and respect, and with this feeling comes a sense of responsibility.

Kinship may also help us to realize that inheritance is a wonderful privilege, which comes laden with responsibilities. Accepting the privileges of inheritance without the courage of taking responsibility for its sustenance is a weakness that we must mutually overcome. Our aim is to strengthen our kinship with nature and to take responsibility in creating a mutually beneficial relationship with it. Once we have this awareness, it is important that we apply this awareness through (meaningful) action to create more balance and harmony around us, not just in our ecological space but also in our social space. This perhaps is the key to our collective well-being and the way toward greater kinship with nature and with humanity.

Healing Forest initiative walks are designed in a manner that they enable us to follow the path of the three core principles of a healthy body and a relaxed mind.

Slow	Calm	Attention
Know	Clarity	Awareness
Grow	Creativity	Answers

Nature mindfulness, forest art, sense walks, and a range of nature games and activities enable us to interact with nature in a manner that helps us to connect deeply and meaningfully with nature, with ourselves, and our society at large.[3]

Healing Forest walks also try to address issues within our value systems, focusing on a range of emotional, mental, and social issues that can enable us to learn important life lessons of kindness, generosity, oneness, interconnectedness, and impermanence. The eventual goal is to create a community of conscious people who can collectively build a foundation based on values of mutual respect and understanding for all life—the reclamation of humaneness in our lives.

One of the facts we all know about nature is that it loves diversity. The more diverse an ecosystem, the more robust it is. There is always space for everyone. A blade of grass, a flower springing out beside a trail, a grove of trees, a puff of moss, a bed of fallen leaves, spiders clinging to bushes and weaving their webs, a woodpecker hammering a tree, acorns on the ground, pine nuts and pikas, butterflies fluttering around, a soft lush meadow under the sun, someone taking a nap on that meadow—it all comes together to become a forest ecosystem. The magic of unity comes from diversity not uniformity.

Humans, however, have slowly drifted away from immersion in diversity into a paradigm of monocultures. Diversity is integral to nature, and the messages we receive from nature speak of how every species has a place and a significant contribution to make in maintaining the overall health of the ecosystem. Nature is self-sufficient and bountiful, as long as we don't tamper too much with its design. Our ability to manipulate nature is both our blessing and our curse. Without knowing our place in nature, we are bound to tamper with nature's design in a manner that might not always be in our best interest. The objective is to create a deeper understanding within us to help us realize our place in nature, so that instead of tampering, we complement its design by applying our imagination, intuition, and intellect for mutual benefit.

We are all part of nature, and yet there is a visible disconnect that we witness in the way we interact with nature. We know that we are all akin to being kin, yet in enacting that knowledge we seem to be floundering.

To understand this disconnect with nature, we have merely to look at the history of the past two hundred years. With growing urbanization and increased use of technology, our interaction with nature has slowly dwindled. This disconnect with nature and the overdependence on industrialization, urbanization, and technology have caused serious physical, mental, emotional, and social health issues.

If we look at human wellness as an indicator, we are at a time in history when we have managed to make some headway in accomplishing the goals of becoming a "welfare state" but have moved further and further away from being a "wellness state." Yet the source of our well-being is fairly obvious and simple: meaningful relationships, especially with nature. The clarity that nature provides us can enable us to find answers to life's challenges. One of the greatest lessons that nature teaches us is how we can form meaningful and healthy relationships.

Can kinship with nature also hold the answer to some of these challenges, especially social and ecological challenges that we face on our planet? Humanity has been collectively striving toward creating more "wellness" in our lives. Our endeavors focus mostly on the need to reduce suffering for human beings, be it emotional, psychological, spiritual, physical, financial, or social. Whether through science and technology, spirituality and religion, politics and governance, the objective is to improve the quality of life and make us more happy, secure, and at peace. Yet that peace and happiness remain elusive. One of the reasons is that material or spiritual wellness is rooted in nature wellness.

I believe our inability to bring about a state of wellness for humanity is tied to our limited humanistic approach. Unless we begin to look at humans as part of nature and work backward from nature wellness to human wellness, the overall goal of wellness will continue to be elusive. Our task is to find the middle ground of human and nature wellness, and to work toward bridging the gaps. If reducing suffering is our goal, that goal has to be holistic, and we need to find a way to reduce suffering for the planet and its species collectively, of which we are a part.

What are some of the major wellness challenges that we face collectively? The issues pertaining to human wellness revolve around equality, equity, social justice, sustainable growth, and health, among a plethora of other issues. Over the past few decades, we have witnessed a steady rise in lifestyle-related diseases: depression, anxiety, stress, insecurity, and other mental health challenges. This is an ample indicator of our failure to accomplish our wellness goals. Moreover, there are innumerable challenges pertaining to nature, which are visible in issues like climate change, deforestation, desertification, biodiversity loss, water pollution, air pollution, waste disposal, and so on. The challenge we face is the dichotomy of nature and human wellness. When we speak of wellness, it has to be holistic. If it remains solely bound to human wellness, we are restricting our own ability to heal ourselves, because our own healing depends significantly on nature. So the goal is simple: helping humans heal, helping nature heal.

Scientific research and technological applications have been instrumental in addressing some ecological challenges. What is still missing, at large scales, is the integration of humaneness into conservation and sustainability initiatives.

The idea of kinship with nature is not new, and if we manage to reconnect humanity with nature, we may need to focus on solutions with a more humane face. Our disconnect with nature is a recent phenomenon and one that can still be addressed. The goal is to understand the need for reconnecting and the tools available to reconnect. Our need to preserve the basis of our collective well-being could be integrated with the idea of fighting for what we love. Ask any person this: if your home was on fire and your children or parents were caught in it, would you flee or rush into the fire to save them? Some battles are worth fighting for, irrespective of the odds.

Let us for a moment imagine that we all have a tree friend. We go out to meet this friend of ours every now and then, as we would like to meet with any other friend of ours. We greet it when we meet it. We touch it and embrace it. We sit and talk to it, and share our joys and sorrows. We hear its songs as the breeze passes through its leaves. We also hear its creaks and cracks, witness its withering, and connect with its roots. We share its fruits and spread its seeds. We find it in us to love it a little. Imagine a world where all of us have a friend in nature—tree, plant, flower, rock, river, mountain, clouds, blue skies, a constellation, the moon—just a friend in nature to keep us connected with our roots. A friend in nature also connects us to some of the core values of love, respect, community, and kinship that help us to live more harmoniously. Imagine, every now and then, we met friends of friends and create an ever-widening circle of nature friends. Imagine a festival of friends, a gathering of friends, a school of friendship, a celebration of kinship. Imagine!

NOTES

1. T. Ryan Gregory, "Evolution as Fact, Theory, and Path," *Evolution: Education and Outreach* 1 (2008): 46–52, https://doi.org/10.1007/s12052-007-0001-z.
2. E. O. Wilson, "Biophilia and the Conservation Ethic," in *The Biophila Hypothesis*, ed. Stephen R. Kellert and E. O. Wilson (Washington, DC: Island Press, 1993), 31.
3. Further information can be found at the Healing Forest website, https://healingforest.org/learn/.

A PEDAGOGY OF HUMAN DIGNITY AND CLASSROOM KINSHIP

Anthony Zaragoza and María Isabel Morales

"Sustainability is in our culture," the Lakota elder said as the students listened intently to a classroom panel focused on themes of culture and sustainability. A P'urhépecha indigenous elder told stories about gardening and reconnecting the community with traditional food knowledge. Another elder offered a slow-paced story about his community's struggle to grow hemp in their lands. After the panel, our class gathered, as we typically do, to process the event. Some students expressed feeling gratitude, hope, and love. A few students offered a critique that led into a challenging discussion. One student shared that the event was too anecdotal and not analytical enough.

Although the stories and poetry offered critical perspectives of sustainability, these students found storytelling a questionable method for a classroom environment that needed to maintain a certain level of "rigor." What makes a classroom discussion "rigorous" or "valuable"? Does the personal not belong in the classroom space? The panelists spoke to the continued impact of colonialism and invited the student audience to reflect and to act. But their messages were delivered via stories that made us laugh, sigh, and dream. Is it possible to bridge the political and the personal? The personal and the structural? Is the classroom an ideal space for this intersection?

Intersectional feminists planted the seed of "the personal is political" to refer to the ways sociocultural issues experienced in our personal lives are intricately connected to politics. The vignette that opens this essay sheds light on the complexity of interweaving the various strands of our living, learning, and acting in the world. As educators, we have found it imperative to braid structural analysis with personal stories.

Kinship is built when we listen, humanize, empathize, and offer mutual support. Solidarity means learning to fight alongside one another, so it must also mean knowing one another's pain and knowing that, while we struggle in different ways, we can still be in relationship with one another. We recognize that we each come from multiple social positions (race, gender, class, sexuality, language, and citizenship status), and such intersections inform our perspectives, biases, and calls for action. Our social positions are differences, not boundaries that prevent our talking, working, loving, and being together. We all have a right to live with dignity and have what we need. Kinship refuses the ideologies of individualism that lead to conformity, isolation, and suspicion of the Other.

Building kinship in the classroom begins with a commitment to knowing students, their communities, families, and epistemologies. For many students, it is important to begin in the personal—with activities that invite sharing about themselves—in order to engage in analysis. In our classrooms, we share stories with one another to build empathy and to build bridges within and across our multiple worlds. In this essay, we braid our own stories with analysis to try to build kinship with our readers and, in doing so, provide a model for what can happen in the classroom when structural analysis is grounded and braided with our students' lived experiences as well as our own.

Political Economic Contexts: Borders in the Classroom

For significant portions of my early life, I grew up in a household with four working-class generations: great-grandfather, grandparents, dad and aunt, and me. We frequently had Sunday family gatherings with nearby extended family. This is no longer possible. An economic stability that came from mill jobs is gone, and the family is spread across the country. Deep, intergenerational ties among families have become stretched and frayed, and fewer and fewer people are able to be close with extended family beyond holidays and reunions. Though at times this may seem like a relief, we know it's not good for us. And that relief may be because we have lost the skills, patience, time, and even inclination to hang in there with each other, especially as the economy and society become more polarized and alienating. Even family has become disposable and replaceable. It's easier to "cancel" people and move on. —Anthony

Divide and conquer is an old tool. Although our natural inclination is to kinship and class solidarity, capitalism promotes competition and division among people and communities by using nationalism and other forms of tribalism. In Anthony's case, the economic instability rippled through his family, and folks were pushed to pursue economic opportunities elsewhere.

The neoliberal phase of capitalism takes divide and conquer to a new level as "ownership-society" ideologies undermine and dismantle the already-limited concept of public good and transform industrial capitalism and its welfare state into a personal-responsibility market state. This is a transformation that suffocates possibilities for broader kinship and encourages atomization. Economically, it looked like this: in the 1970s, a growing capitalist class consensus emerged to embrace efforts to transform and downsize regulatory control, to march toward increasing privatization and government support of corporations, to cut social spending such as financial aid, public housing, and welfare—and this resulted in an even more frayed and overstretched social safety net. This made budgetary room for a series of tax cuts for the wealthiest families and corporations, and a series of further increases in military, police, and prison spending.

Five decades of neoliberalism have led to severe concentrations of wealth and an accompanying growth of inequality and various social maladies like homelessness, addiction, and decreases in quality of life and life expectancy. We see it in the gentrification rippling out from wealthy hubs, sending rents skyrocketing and leading many to move farther from work or to live without housing. Construction cranes pop up to build mixed-use condos for tax-dodging developers as homeless camps pop up in public parks. The camps get cleared for profit by companies who throw in the dumpster what the scattered houseless can't carry under the watchful eye of publicly funded police. Policing, mass incarceration, and surveillance maintain the unstable and unequal social system. Data-mining and public relations corporations commodify our fears and insecurities, our needs and desires, even our searches for kinship.[1] Meanwhile, we are heading for global heating, extreme weather, and climate chaos as extractive industries are unleashed and defended. Our students and their families cannot but be directly and indirectly impacted.

As neoliberal policies have increased inequality, concentrated wealth is made to seem normal, natural, and attainable. Neoliberal ideologies don't stop at the classroom door; in fact, they've been perpetuated in higher education. The ruling class is experienced at convincing middle-class folks that the problem is due to "those homeless," "those immigrants," "those thugs," or "that country." This sort of scapegoating happens in our classrooms, especially from students with less proximity to the newly arrived, people of color, and the poor of all colors. In our classrooms, we bear witness to the ways white supremacy and capitalism have an impact on our student's lives and the comfortable assumptions they often make about one another. Meanwhile, ruling elites remain unseen and unaccountable. In these neoliberal times, collective power is undermined and the exploitability of working-class people increases.

Transgressing Borders

"Bridging," writes Gloria Anzaldúa, "is the work of opening the gate to the stranger. . . . To bridge means loosening our borders, not closing off to others."[2] Of course, we must honor our different identities, access to resources, and cultural backgrounds, and recognize that many times being with others with similar stories helps us to heal and thrive. It is also true that we are not just *in* the world; we are also *with* the world, and *with* other people. As such, it is important to build bridges across the white supremacist, neoliberal, capitalist patriarchy—to echo the author and social activist bell hooks.[3] In our classrooms, "bridging" is essential and moves us to build connections, in the form of coalitions, across and with(in) communities to challenge isolated individuality.

Knowing the history of our common struggles opens possibilities for empathy and earnestness among different groups that can bridge the distances between our different identities and experiences. At the same time, some identities and experiences have been marginalized or presented as deficits. In our classrooms, we seek to name students' experiences as "community cultural wealth"—the many forms of wealth our communities have to build on.[4] Our students have a wealth of knowledge about navigating complex economic, cultural, linguistic, and social systems in their neighborhoods.

This concept of community cultural wealth can help provide strength and set the stage for transgressing borders. All individuals must draw on something from their cultural capital in difficult moments when faced with the gravity of the white supremacist, capitalist patriarchy.[5]

Collaborative Learning and Co-Constructing Knowledge

We're in the heart of the traditionally Black, working-class Hilltop neighborhood currently undergoing deep and rapid gentrification. Many of our students are nearby students of color, nearly all poor and working class, and roughly two-thirds women. Our students are usually returning after having had some disconnection from school while on another path for years or decades. Often schools have failed them.

On the final day of fall quarter, we gather to read memoirs. Juniors work all quarter on drafting, revising, and editing a story from their lives to share with peers. Seniors know what's coming; they've been through this. There's great anticipation as well as nervousness around this deep sharing. I share, too. And I know the tears that will fall from most, if not all. Students share about lost loved ones, births of children, domestic violence, recovery, near-death experiences, comical scenes, breakthroughs, falls from grace, encounters with police, time in prison, falling in love. The memoirs offer a spectrum of joys, traumas, and motivations that are common to our students and the communities we come from. The Evergreen Tacoma memoir assignment builds camaraderie.

With these memoirs, we honor the whole student, our humanity, and share from our lives. Here we can recognize possibilities for kinship and opportunities to develop it. This kinship will be crucial during the educational journey ahead as it offers a gravitational force keeping people connected, closer to the learning and returning to school. Because we're a "village" learning community, kinship is essential as we learn, disagree, and cohabit our learning spaces. The memoir assignment and the kinship it encourages were gifts given to us by our founder-director Dr. Maxine Mimms and second director Dr. W. Joye Hardiman, two brilliant Black women educators. —Anthony

Reflective writing is a challenging but necessary part of consciousness raising and kinship building. The act of writing our stories is an "endless cycle of making it worse, making it better, but always making meaning out of the experience, whatever it may be."[6] Sharing memoirs, or *testimonios*—a form of narrative that has been historically used as a tool to illuminate sociopolitical struggles and voices of marginalized peoples—helps to break down essentialist and dehumanizing categories that serve as technologies of control. When our students courageously share their testimonios, we are all reminded that when we dare to be vulnerable with one another, walls disappear.

In our personal histories, we've probably all had low moments, near surrenders, and unforeseen relapses we fought back from.

> *I remember walking with my brother away from my mother's house after his addiction to crack led him to destroy their relationship. His holey, rickety suitcase didn't zip. His life, or what was left of it, was in it. No matter which way he held it, something ended up on the rocks and broken glass of the alley. I was in my fourth year of college, and I could think of nothing to help my brother as he wandered addicted into those glue-trap streets. I felt helpless and scared as I watched him get farther and smaller as he wrestled with his suitcase to keep the rest of his life from falling out. I felt desperate, angry, and guilty; I could do nothing but witness and empathize. I felt that emptiness and loss in my heart, but it would be years before I could, as [bell] hooks has described, "understand the core, underlying truths, not simply that superficial truth that may be most obviously visible."⁷ It's too easy, even lazy, to simply tack it up to his bad choices. We must understand the social, political, and economic forces that led to this scene: the gutting of our industrial economy of Northwest Indiana, the drugs (PCP, crack, and crank) flooding the region as replacement economies, and the many histories of conquerors deploying drugs and alcohol against us. In Northwest Indiana, the boom was bust, and the social framework was bulldozed. The salvageable parts were shipped south, and people were scattered, jailed, drugged, or forged on; some kept up the fight for transformation. —Anthony*

When we, students and faculty, share our stories with honesty, courage, humility, and love, we collectively build on one another's political dreams of a more just world for all. By opening ourselves to students, students may decide to open themselves to us and the learning community. The borders we, as educators, bear witness to in our classrooms are multiple. We have student veterans struggling with post-traumatic stress disorder, mental health, food and housing insecurity; immigrant students stressed because of their status; and nonbinary students navigating heart-breaking isolation, among other issues. We as working-class faculty also experience our own indignation for the injustices our families experience, and it is humanizing when we are honest with our students about this.

Building Empathy and Becoming Kin

I loved my first-grade teacher, Mrs. Johnson. I remember her dark black curly hair, her tender voice and her strong demeanor. I loved her classroom. I loved the smell of crayons fresh out of the Crayola box and the sweet fruitiness of Mr. Scent markers. To this day, these smells spark deep joy in me.

The day that Mrs. Johnson invited my mamá and I over to her house for "chocolate con pan" is one of the most special memories I have. This was a pivotal relationship-building moment between my teacher and me. Most importantly, between my teacher and family. These memories remain alive and vivid in my heart—I still remember how excited I was when my teacher added marshmallows to my hot chocolate! I hold on to these memories just like my mom still holds on to the crochet blanket Mrs. Johnson gifted her that day. In retrospect, while I may remember events and activities, I mostly remember how I felt in her classroom—welcomed and appreciated. I was acknowledged as an important and whole being. I felt respected as the older sister that at times needed to go to her preschool brother's classroom to cheer him up. I felt loved. I felt like every part of who I was came with me into the classroom space. Mrs. Johnson was the "warm-demander" that the African American educator Lisa Delpit describes who "expects a great deal of their students, [and] convinces them of their own brilliance."[8] Mrs. Johnson made sure to let me know that I was brilliant—that my immigration story and bilingualism were not deficits but rather an important part of my community cultural wealth.[9] —María Isabel

These sorts of reflections are important as we think through kinship building in the classroom. We remember our Sunday gatherings with our family before the economic instability scattered us. We remember the individuals in our lives who made us *feel* part of a community as whole human beings. We remember our own migrant dreams. Some students experienced great strife as children and others had important examples that strengthened a sense of self. One activity asks students to remember who they were as children by using art. This activity opens our hearts to one another as beings with multiple-layered histories, and we find that we have more in common than assumed. When we bear witness to one another, we plant a seed of empathy—the willingness to understand other people's feelings, life experiences, and perspectives, *and* the openness to vulnerability this requires. Empathy is nurtured. Kinship is cultivated.

Empathy is a value that all of us can work to grow within ourselves and with one another. Just about everyone we pass has their own struggles. Our colleague Gilda Sheppard talks about our Tacoma students as "walking miracles," because of all the things they endure and fight through to get their education. We must be empathetic both with the struggles and with the resilience, resistance, and refusal to give in. We need both.

Empathy begins with learning how to *listen*. We have to be present with one another and deeply listen. Listening is an act of humility and matters in the classroom because "listening to all that come to us, regardless of their intellectual level, is a human duty and reveals an identification with democracy and not with elitism."[10] To listen is to be consciously engaged in ongoing reflection with our stories and open to the possibility of changing our perspective and relation to the world. Stories are a source of power and resilience, not just as individuals but as communities and agents of history. They give important historical perspectives that add depth to our learning of political or economic structures.

Conclusion

> *"Hoy por ti, mañana por mi" is a phrase I grew up hearing from my mother. This philosophy is practiced in the packing factory she works in. When there is a death or sickness—whether someone directly tied to the packing factory or a relative—the women call a quermes (potluck) to fundraise money for the person or family affected. The women organize the event and the whole factory supports it, purchasing food or donating for supplies. Recently, a coworker was diagnosed with a sickness that forced her to stop working. The woman is undocumented and as such has no health insurance. Her expenses were unmanageable. My mother and her coworkers came together to fundraise and gifted her with cash and a basket of food. It can take a lot of planning and energy to cook (or purchase whatever is needed). For people who work in strenuous conditions, it is deeply powerful that they are still willing to do what they can to help the community. It is the way it has to be, says my mother, because today they are in need, but tomorrow that person in need might be us. —María Isabel*

This story reflects one of the underlying goals of teaching the way we do—to help students see that we thrive when we support one another as community. "Hoy por ti, mañana por mi" is a praxis of reciprocity and a counternarrative to the anti-immigrant rhetoric that portrays immigrants either as criminals or solely as victims and not agents. These family stories feed our own dreams of kinship and love as educators. Our commitment to weave our lessons with values of reciprocity and community care is part of a pedagogy of kinship that allows us to "open our doors to the stranger" to learn together and move toward communal, national, and international cooperation.

While community solidarity is important, it also takes an incredible amount of work and various forms of labor (emotional, organizational, tracking, and communal management). It requires intentional attention to neurodiversity and cultural diversity, for example. All this—in addition to the reading, writing, listening, speaking, planning, and presenting—is part of academic teaching (not to mention research). Often, this kinship work is unsupported, untrained, uncredited, and not widely discussed in institutions of higher learning. But it is a key part of our jobs as teachers, learners, and scholars, because in humanizing the learner, in validating their voices and experiences, our own voices and stories are humanized. To work from a pedagogy of kinship, we move toward a community of respect, equity, inclusion, care, love, and justice.

NOTES

1. Emma Briant, *Propaganda Machine: Inside Cambridge Analytica and the Digital Influence Industry* (London: Bloomsbury Publishing, 2020).
2. Gloria Anzaldúa and AnaLouise Keating, *This Bridge We Call Home: Radical Visions for Transformation* (New York: Taylor and Francis, 2013), 3.
3. bell hooks, *Teaching Critical Thinking: Practical Wisdom* (London: Routledge, 2010).
4. Tara J. Yosso, "Whose Culture Has Capital? A Critical Race Theory Discussion of Community Cultural Wealth," *Race Ethnicity and Education* 8, no. 1 (2005): 69–91.
5. hooks, *Teaching Critical Thinking*.
6. Gloria Anzaldúa, *Borderlands/La frontera* (San Francisco: Aunt Lute Books, 1999), 73.
7. Anzaldúa.
8. Lisa D. Delpit, *Multiplication Is for White People: Raising Expectations for Other People's Children* (New York: New Press, 2013), 70.
9. Yosso, "Whose Culture Has Capital?"
10. Paulo Freire, *Teachers as Cultural Workers: Letters to Those Who Dare Teach* (London: Routledge, 2018), 71–72.

THE VOCATION OF CARE
Alison Hawthorne Deming

—*Archibald MacLeish Field Station, Northampton, Massachusetts*

Someone has tied dozens of stones
        to the limbs
    of the London plane tree

white string glued to rocks
        clipped with clothes pins
    forcing the limbs to fountain

down toward the bed of mulch.
        I can't tell if the work is
    science or art though it makes me

think of the Rene Dubos book
        The Wooing of Earth.
    Spindly chestnut trees

sprout in the seed orchard
        an experiment
    blending bacteria

into its genome to quell
        the toxin that felled
    the forest giants of their ancestry.

It makes me think
>   (so strong is the kinship
>   between mind and nature)

of my grandfather, taciturn
>   Yankee doctor who
>   preferred plants to people.

Luther Burbank type. Little pot of hot wax
>   tied to his belt
>   as he grafted his fruit trees

to see if pears and apples
>   might be coaxed
>   from the same tree.

His doctor office sat next door
>   to a water-powered mill.
>   Connecticut industry

ran on water. Clockmakers
>   hatmakers, axe and gun makers.
>   He must have treated

arms yanked out of joint
>   on a fly wheel, fingers
>   lost to blade or gear. Worse.

I picture him walking
>   the woods of my childhood
>   with my father, the two men

gentling the leaves of trees
        they knew would die.
    He'd have been happy

to see the new science
        meet up with the old
    here at the field station

where a closet full of
        off-duty microscopes
    sit lined up on a shelf

each wearing a hood
        labeled Leica, cloistered
    like monks in silent prayer.

THE INVITATION
Jill Riddell

What I was doing was the opposite of vandalism. If caught, I wouldn't be arrested, I assured myself. And I didn't plan on being caught.

I opened the back door of the car and pulled out the bag with flashlight, duct tape, scissors, and box cutter. I slung it over my shoulder. I'd thought ahead enough not to have worn a coat made of slippery fabric but instead wore a sweater and wool jacket so the tool bag wouldn't slip from my shoulder. I walked around to the trunk. Inside was a piece of black foamcore I'd bought at an art supply store, laid out in my basement, and painted on with white letters. The sign was so huge that, to squeeze it in, I'd had to lay down the car's back seat.

It wasn't late, only nine. But this was November and the park was completely dark. Streetlights cast sufficient glow that I didn't need the flashlight. I looked around, scanning to see who might potentially try to interfere. Across the way near an apartment building, I detected the movement of two people walking, but they were heading the opposite direction. Behind me, nothing, no one. Ahead, along the paved bike path and farther out into the grassy lawn, all was still.

A few days earlier, I'd been at this park with a friend, Erika Dudley. Promontory Point is a place we both love. Its peninsula juts out gently into Lake Michigan, and when I was pregnant I walked its circular path counterclockwise every day. For a long time, Erika has been coming here regularly to watch the sun rise.

A couple years ago, Erika convened a small group of Chicago women to start discussing the built environment. Once a month, we gather at the Hyde Park Art Center, where we talk about all kinds of things, mostly circling around how public spaces in Chicago can become more inclusive and welcoming. There's a disparity between how white people and how Black and Brown people use public spaces and how safe we feel. When we're together, the five of us are not exactly activists, but more like observers who bring to meetings what we've seen or heard over the past month. We listen to one another's discoveries—the good, the bad, the in-between— and sit with it. We talk about it. Sometimes we even nap about it, laying our heads on the table, on top of crossed arms, and softly closing our eyes.

I'd brought Erika here to see something specific. Near the entrance of this wonderful park were twenty-seven signs, but instead of a welcome, each issued a gripe about visitor behavior: "Pay for Parking at Pay Box," "No Loitering, No Trespassing," "No Alcoholic Beverages."

"This one has got to be the worst, though," I said. On a chain-link fence by the parking lot was a series of signs with "arrest" emblazoned on them in big red letters. Together, we counted. It was an even dozen. All had the same message: "No Alcoholic Beverages Allowed on Park District Property—Violators Subject to Arrest."

"Clever the way they managed to work the word *violator* in there, too," Erika said.

We walked into the park a bit farther. The signs calmed down a bit—info on the park closing at eleven, the need to curb your dog, the imperative to dismount from a bike—not unfriendly exactly, but not welcoming. Each was a correction of potential mistakes we might make. On the way out, I pointed to one large, well-made metal sign holder. It looked old. The size and shape of it wasn't like the other signs—it was horizontal, wider than high, and had two posts sunk deep into the ground. Although it had a grand sense of permanence, inside the frame was nothing. It was absolutely empty, except for the strong north wind sweeping through the gap.

"Once upon a time, this must have been the official entrance sign," I said. "The one announcing that this was Promontory Point."

Nothing had been inside the frame for many years.

I'm someone who, in an art museum, has to force myself to look at the art before I read the label. I love the written word that much. If I go to a natural area, I'm going to stop first at the kiosk, the one with interpretive materials hung behind a sheet of plexiglass. I like to know what's what.

Usually, though, the kiosk is the Bureaucratic Dumping Ground of Dispiriting Treatises. An eight-and-a-half-by-eleven inch sheet of office copy paper talking about some terrible wasting syndrome white-tailed deer are suffering from. A warning about mosquitoes carrying eastern equine encephalitis. The kiosk highlights threats without any counterbalance of encouraging words of welcome.

Is a landscape with germy mosquitoes and sick deer one I want to explore and develop a sense of kinship with? Maybe, because I already love being outside. But if I hadn't had a history of positive experiences before this, if I already felt myself on unfamiliar ground, my interpretation of those signs would be: "I don't want to be your friend. Stop bothering me and go away."

I do realize that Midwestern parks are underfunded, that no agency has a giant communications and marketing staff. When I served a term on the Illinois Nature Preserves Commission, each year the General Assembly swept the commission's entire budget into the general fund to cover the state's debt. Over and over. When an agency is starving, niceties like signs are hard to make room for. It makes me wonder whether park employees feeling down on their luck—perpetually underfunded, their efforts unheralded—gloomily put up grouchy signs as some sort of revenge.

The napping part of our built environment meetings was a recent, thrilling addition. Already, the meetings had been a bit magical, made that way both by their open-endedness—we didn't *do* anything exactly, and we were mostly there to *be*—and by Erika's food. She's a chef who trained in Paris at Le Cordon Bleu, and each time we meet, she brings something to eat, or, as she puts it, something for us to take pleasure in.

The beauty of the food was all Erika, but the napping idea originated with someone else, an artist from Erika's hometown of Atlanta. Tricia Hersey started the Nap Ministry to suggest that rest can be a form of restorative justice. Women, especially Black women, require rest to give them the strength to resist all the "'isms." Women who were enslaved could not control when they worked and when they rested. Hersey's Nap Ministry advocates for women to reclaim their bodies and their time and to make napping part of their practice. They should take long naps—*because they can.*

Perhaps now is the time to mention my own whiteness and the generosity of my having been included in the built environment group, given that, other than me, the group consists only of Black and Brown women.

THE INVITATION

I've been thinking about the people who love natural areas, parks, and open spaces—those of us who feel confident being in them—and who are concerned that, in the future, there may not be so many voters advocating for their protection. We do a lot of promotion of the outdoors by speaking to how it relates to physical fitness. But what if we promoted parks as places to rest? Not to meditate or be mindful—I mean, go for that if you want—but to sleep. Soundly . . . complete with dreams . . . with the sunlight flickering overhead through the leaves of trees.

"What if we had a 'nurturer-in-residence,'" Erika said at one of our meetings. In context, she meant for someplace like the Hyde Park Art Center. It already sponsors artists-in-residence. But quickly, we started throwing out ideas of other places that could benefit from a nurturer-in-residence.

In a park, it could be someone who personally would welcome a visitor. The nurturer-in-residence would offer a sleeping pad, a blanket, a pillow, a small cup of chamomile tea, and an exceptionally well-made brownie. As the visitor dozed off, the nurturer-in-residence would keep an eye on their belongings and make sure no harm came.

The nurturer-in-residence could provide bandages if someone tripped and skinned a knee; advice on where to see an interesting bird; directions toward edible violets that the visitor might gather and later, back at home, sprinkle onto a salad. If an elderly person arrived, the nurturer-in-residence might sit with them on a bench and deeply listen.

Strangers gravitate toward children—toward noticing them, smiling at them, stepping in to offer care when it looks like it's needed. To nurture them. But what about the seventy years of a human life span spent *not* as a child? Who gravitates toward us then? A nurturer-in-residence might.

After Erika and I looked at the signs, we got back in my car. This particular November was prematurely cold. We turned on the seat warmers.

"I'm thinking nature needs to be more like a dinner party where you invite people a bunch of different ways," I said. "You know how it really isn't enough anymore to send a single invitation for a party, you need to reach out several times, in different mediums. And it helps to create some anticipation, and when they arrive, you delight them with different little things that are amazing."

"That's it," Erika says. "I host events and if someone doesn't show up, they're not off the list because they didn't come. The invitation will be extended again and again."

"So the question here is, what invitation does anyone receive to encourage them to come here, to a park?"

"The park is its own invitation, don't you think? Like that tree over there—look how inviting that tree is." Erika pointed through the windshield toward our left.

A small grove of trees was partially blocking the view of Lake Shore Drive, a big road that stretches between this part of the park and the scenic peninsula. "Which tree? The one with all its leaves still on or the one without?"

"With the leaves. It's like ivory. Pale, pale yellowish white. I've been looking at it as we talk. I think, 'Oh, that's so beautiful, isn't that so beautiful,' and I see cars driving by and I like to think that maybe it catches a driver's eye and they're thinking, too, 'Isn't that wonderful?'"

I look where Erika is pointing and it's one of those things where something wild looks not like it's catching the light but making its own—a swamp white oak, lit up, still hanging on to a robust crop of leaves, each one shimmering.

"While we've been out here," Erika said, "I've been thinking of African Americans' relationship to the natural world—in the United States, with slavery, with all the years that have come and gone since slavery, and our not feeling welcome and not feeling that one has the privilege to be present in public spaces."

Erika's gaze remains on the tree. "Especially lately, with social media it's documented—Black folks out barbecuing, having a family reunion in a state park, and they're attacked because they're in a space where other people think they're not supposed to be. And what does it mean to feel like you're not welcome in a public park and you have to be looking over your shoulder for something to go down. Or to be worried that someone else is going to be suspicious that you might be the instigator of something. How does one move through an open space with the knowledge that, at any given moment, something dangerous might happen?"

"And it takes so little to scare a person, to keep someone away from doing something or trying something," I said.

"Exactly."

"I'm thinking now about our kids." Erika's son and my daughter went through fourteen years of school together, nursery through high school. "You know how when someone is harsh with a child—they might say something mean or rough to them and, after, you can't win back the child's trust by saying 'Sorry, let's move on.' It actually takes a super amount of coaxing. It takes a hundred acts of kindness to prove to that young person that you have their back."

"One bad moment and to compensate, there have to be demonstrations of the opposite, of goodness. Maybe a hundred. Maybe a thousand," Erika said.

Large, awkward leaves of sycamore trees were blowing in front of the car, tumbling, flying, scraping along the ground at tremendous speed.

"I've had conversations about fear and trauma with Black folks—you're never ultimately a first-class citizen of this country," Erika said. "We tend to think that nature is for everyone. And if someone doesn't accept this, doesn't agree—we say, 'You say you can't afford to go to the Grand Canyon, but, hey, this city park is right here and you're not taking advantage of it. You're not participating in this, and this is free! And it's just down the street.' As if proximity means accessibility, as if there being no financial cost means you can do it, that you ought to *want* to do it. The burden again falls on you to change your mind and decide you're welcome, and not on others to see that *why* you might feel unwelcome is generational, long standing."

The next day, I went back with a tape measure to check the dimensions of that metal frame where some sort of sign had once stood. Then it was a few days later and I was back again. I picked up my unapproved replacement and hurried into the empty park.

I hadn't counted on such a strong wind and how a piece of lightweight foamcore would catch it. This November was crazy. I was being buffeted along as if the sign were a sail.

When I reached the frame, the sign I made turned out to be too wide, so I used the box cutter to trim it some more and then—*darn it*—made it a bit too small. The duct tape didn't want to stick because the metal was cold. The whole thing was a struggle and I was nervous.

In the book *Braiding Sweetgrass*, Robin Wall Kimmerer says that when she has a new group of students the first thing she tries to convey to them is—and I'm paraphrasing here—that Nature doesn't hate our guts. That Nature wants us around, that it cares about us somehow. When I was back in my basement deciding what to write, I held in mind that idea. I ended up painting, "WELCOME! It would have been so easy to stay indoors. But you made it here! Way to go! Love, The Park."

And on the other side: "THANKS FOR BEING HERE! We're so glad you came to see us. Please come often. With much love, from the Lake, the Trees, the Grass, the Sky."

Making sure people have space to rest and offering visitors kind words aren't everything. They don't counteract trauma inherited from centuries of slavery and deep-seated racism—or compensate for harms being inflicted in the present. Wariness can't be converted instantaneously into kinship. I finally got my sign installed, knowing it couldn't really coax anyone into a public space where who knows what might happen.

Here's what I think the deal is, though: You could lay the same criticism on anything. Rest and kindness *aren't* everything, but nothing else is "everything," either. No single act fixes "everything." But it might make one small part of one problem a fraction of a percentage better than it was before.

If we need to perform a thousand acts of kindness, best get started. One is better than zero.[1]

NOTES

1. Sources and inspirations: Erika Dudley wears many hats, but her professional home is at the University of Chicago, where she works in the Civic Knowledge Project within the university's Department of Civic Engagement. For more, see Brooke Nagler, "A Chef's Approach to Community Building," *UChicago Magazine*, September 21, 2018, https://mag.uchicago.edu/arts-humanities/chefs-approach-community-building#; and Vocalo Radio, "UChicago Organizer Erika Dudley Says Chicago Has Radical Imagination," *Vocalo*, January 28, 2020, https://vocalo.org/erika-dudley/. On Facebook, look for The Loom/Progressive Conversations on Food, Arts, and Culture. Tricia Hersey's Nap Ministry can be found here: https://thenapministry.wordpress.com. To learn more about Promontory Point, there's this: https://www.promontorypoint.org. The Hyde Park Art Center, which provides a home where the Built Environment group meets each month, is at https://www.hydepark-art.org. These books have been helpful to my thinking and practice: Carolyn Finney, *Black Faces, White Spaces: Reimagining the Relationship of African Americans and the Great Outdoors* (Chapel Hill, NC: University of North Carolina Press, 2014); and Robin Wall Kimmerer, *Braiding Sweetgrass: Indigenous Wisdom, Scientific Knowledge and the Teachings of Plants* (Minneapolis: Milkweed Editions, 2013).

EPILOGUE—ATTENTION, CURIOSITY, PLAY, GRATITUDE: PRACTICES OF KINSHIP

John Hausdoerffer, Robin Wall Kimmerer, Sharon Blackie, Enrique Salmón, Orrin Williams, María Isabel Morales

Figure 1. Zoom screenshot of the conversation partners.

January 15, 2021: Kinship's *coeditor and coauthor John Hausdoerffer invited one author from each of the five* Kinship *volumes for a hopeful discussion. Pictured above are the participants: Sharon Blackie (top left, from* Partners*); María Isabel Morales (top right, from* Practice*); Enrique Salmón (bottom right, from* Place*); Orrin Williams (bottom center, from* Persons*); and Kinship's coeditor Robin Kimmerer (bottom left, from* Planet*). We met during a moment in time that felt in need of deep conversation about the practice of kinship. In the midst of the COVID-19 pandemic, during a month in which American deaths from COVID-19 passed four hundred thousand and in which the mental health stress of necessary stay-at-home orders kept* *people apart; nine days after the white supremacist, fascist storming of the US Capitol to violently overturn a fair election; and five days before the inauguration of President Joe Biden and Vice President Kamala Harris—we gathered together via Zoom to reach across public health, cultural, technological, and democratic barriers to imagine a new (or ancient) practice of kinship, together.*

John: Welcome, friends and Kinfolk. Please, simply share whatever comes to mind and heart when you hear the word *kinship*.

Robin: When I think about kinship, I think about a mutual exchange of gifts and responsibilities. I think of *kinship* as a verb—something that we *do*, by each member of the living world having a beautiful gift to share with one another. It is the exchange of those gifts that makes kinship more than ancestry or common history. Kinship is mutual regard and respect for one another. Kinship is that interplay of the gifts that we're given and the responsibility that we have.

John: I really like the idea of *kinship* as a verb, as an active form of interplay. Your book *Braiding Sweetgrass* is so important to this topic. Can you help us think about the "braiding" that happens in our *Kinship* volumes?

Robin: The braiding that has really struck me in *Kinship* is both across cultures and experiences, but also the scale, the scale of the very practical embodied relationship—braiding with soil, with seeds, with insects. I want to say there is a common ground across these beings and scales, but it is not an easy commonality to me. Kinship arises not because we are the same, but because we're different. It is the respect for that difference and the love that comes across difference that really makes kinship hard. It is easier to love those beings who are the same as you. It is a lot harder to love those who are different from you. It is a lot harder to weave love across scale and culture and into the sovereignty of other beings.

John: That reminds me a bit of Gandhi's book *Satyagraha*, in which he says that passive resistance is not so passive; it is not easy. It takes work. Similarly, the word *kinship* might sound gentle and easy, but the way you just described the creation of a commons of love across difference shows that it takes work. We're speaking only days after the January 6, 2021, white supremacist terrorist attack on the US Capitol. Creating a commons right now between Americans, let alone across nationalities and then species, takes work.

Enrique: My name, Tarahípame Chómari, means "Moves along with Deer." I am Rarámuri. English is actually my fourth language. When I think of kinship, I agree totally with Robin. Kinship is an action. It's a verb. It's something that we engage in, not something that we watch from a distance. I think of relationships, and relationships that we have with everything around us as kin, as our direct relatives.

Kinship is actually one of these words that has several meanings and actions, similar to a word like *breath*. For example, *breath* means our spirit, our energy and, you know, if you get into *Star Wars*, like I do, it's the force. George Lucas did a good job describing it. That breath or force is something that everything has. We all share the same breath.

Currently we're sharing this common pandemic, this common disaster, because of the shared breath. So, we can have positive, but also not so positive, thoughts and actions and results from shared breath.

Most American Indians, my people included, believe we were created here. Coyote and Eagle created us, depending on which belief we have, and brought us here for a purpose, as stewards who minister to this ongoing interrelationship that everything shares. My recent book *Iwígara* looked at this kinship in terms of plants. That's really how I was raised—learning all these deep kinship relationships with plants, learned from my mother and grandmother on my mother's side. My grandfather as well, who was an *owéruame,* a singer, which means a healer through song, because the song is brought to life again through shared breath. Kinship is shared as deeply as breath, and has a healing power.

Sharon: Kinship makes me think about two things. There is a kind of reciprocity in the way that we look at each other. For example, we might tend to ask a question like, What is the crow in my mythology? But when I go out there, I'm always thinking, Well, what am I in a crow's mythology? What does the crow see when it looks at me, and just as important, what do I want it to see?

I guess that goes back to what Robin was saying about responsibility. You know, you have a responsibility to be some kind of valid participant in the world in which the crow is as well. It's a quality of mutual attention and mutual respect, as much as I can find a way to put it.

And the other thing is fluidity. In the tradition that I come from, which is very much the Celtic traditions of Britain and Ireland, women, particularly, and the female deities are constantly shapeshifting in and out of animal form. There is no sense that this is a lower form. It's quite the opposite. Sometimes you need the wisdom of a crow or of a fox, because there are lovely old stories about how human wisdom sometimes runs out when you want the answer to a problem you have to solve. So you go to the elders of the village and the elders of the village don't know, and they send you to the older elders and to the next village, and they don't know, and finally they send you to the oldest animals, and the oldest animals always know the answer.

Kinship is respecting the wisdom that we in human bodies don't have but that is available, say, to a crow. There is fluidity in that ability to not really see ourselves as separate enough that we can't, in some way, take on the wisdom of a crow or another animal.

Orrin: Kinship is one thing *and* it is a diversity of things. You can think about it from a familial perspective. Of course, you've got children and grandchildren. And then there's community—there's all of that working for us.

But I resonate with what Enrique and Sharon were saying, and for me kinship also involves a return to indigeneity, coming as we did from the continent of Africa. Then merging into this American scene, even at the level of DNA, I cannot just move forward thinking that I carry only those African genes, right? I carry so many genes. If the stories from my ancestors and my family stories are correct, then I carry the ancestry of most people, everybody on this call as well.

Even the events last week (and I'm not going to spend a whole lot of time on that): kin has been disrupted for all of us. Those people [who attacked the Capitol], they are reacting to being white. The flip side of that is everybody should be proud of their skin, because skin is a significant part of being human. Coming out of the African experience and being able to walk around this planet as we have over the past seventy thousand or eighty thousand years, skin has been a reflection of that history. Skin was also important in the context of what made you be able to survive in whatever environment you were in. So when you look at me, obviously my more recent kinship is of course related to being in a tropical zone—Africa. As I look at other folks on the call, I think, Wow, these folks emerge from that same continent and made that journey. So we should embrace that, how we look, right?

But then kinship becomes a cosmic kind of thing. Going back to what Enrique says about breath or energy or chi—or whatever words you want to use. These electrical embodiments relate to trees and plants. You can get terpenes from marijuana, but you can get terpenes from being in the forest. In the forest, these things go into your lungs and they're healing. When I see you all in person, even the expression of a handshake or a hug, whatever the case might be, that actually connects us energetically. We don't pay a lot of attention to it, but it raises up all kinds of energy. We do that with children. We do it with our pets or animals, our plants, even down to the microbial level—related beneath the surface.

From all of this, people develop relationships to place. Sometimes it's hard for people to move on and go somewhere else. And they're always rooted in place because of this connection. Even with things that are below the surface of the earth, like how trees are connected beneath the surface. And then you also have these cosmic connections. I was reading *The Most Radical Thing You Can Do* and actually got stopped because they were talking about this field that is connected to the rest of the universe as hydrogen atoms, connected to other galaxies, the whole nine yards, and they can witness this.

For obvious reasons, the powers that be try to keep us from regrouping, from reconnecting and noting the fact that we have this field. If you're raised in a church, if you were raised here, you're expected to not reengage these connections that we have.

John: I love the way your essay weaves together what you call "skinship," which I'm hearing as somehow *both* a shared origin and a pride in diverse identities. Can you talk about that weaving?

Orrin: Listening to Sharon, noting these similarities, I am reminded of this movie called *The Forest*. I must have watched it, like, six times. It is about how the British deforested Ireland because the Irish had this whole cultural and spiritual connection to the forest. I was like, "Damn, never thought about it from that perspective." It's just like killing all the buffalo, bringing down people by cutting them from their cultural and spiritual connection.

John: I think part of the practice of kinship is around healing from ancestral trauma. I like how connectivity in your essay calls for healing, like when you say, "Thus, I proclaim that we are cosmic stuff, sourced from the same stuff that forms all actions and manifestations of the Cosmic Mother-Father." I think the healing begins with recognizing that there's a shared wound from the alienation from this "source" you just talked about, across cultures.

María Isabel: I just really am sitting with all the stories and I'm really grateful. I resonate with a lot of what you're all saying in terms of kinship. I think of love as the first thing that comes to me and I think of hope, and not this sort of romanticized notion of hope. I think of hope as sort of an action against fatalism, against this notion that we are powerless and then we can't change and we can't transform. As an immigrant, as someone who has moved from place to place (not just that my generation moved from place to place, but my parents' generation and my grandparents' generation). I was born in Mexico, I grew up in California, and now I live in Washington.

My mom and I crossed the border. My grandparents, especially my grandfather, were guest workers in this country. Movement has been a part of my family stories. The stories I hear from my grandfather about coming to the United States to labor in the Southwest, and then eventually coming to the Pacific Northwest, that movement drives our stories.

Stories have been a really important way for my family to maintain this heart and this connection to the roots. Those roots you talk about, Orrin, for me are those lands that we come from down in the Pacific in Mexico, even though we're not in those lands anymore. The one thing that holds us as family is those stories and nurturing those memories of sustainability and of being. As an immigrant, I think in a generation I will have built more of a home in this place, I will connect to this place and see it as a home. Yet there is still a lot of memory and connection to where I was born, the land of my grandparents, and the connection with these lands we are no longer in.

I think of memories and stories when I think of kinship. Right. Where did I learn the importance of connection with each other? From other people and how people made me feel. Kinship is an action. Kinship is also a feeling of being connected to one another and nurtured. Kinship is a feeling of love of what emerges from *being in relationship*. Kinship is sharing time and space with folks that nurture memories and nurture stories of love.

I reflected on this in what we wrote about kinship in the classroom for the *Kinship* volumes. I thought about the ways that traditional schooling continues to ask many of our students (particularly those that don't represent the dominant narrative told over and over in schools) to check a part of ourselves in the classroom. As someone who has many identities and is still building my own, I feel like I live in this in-between place. Some people talk about it as a kind of borderland—like I'm part of here, I'm part of there, yet I am going to build my own identity.

When you ask somebody to check a part of themselves at the door, that's really dehumanizing. Kinship is when you don't do that, right? We invite every part of who we are, whether it's a classroom space or a conversation. Be intentional about welcoming, about checking in with each other and all that we are.

We are all our different identities, different voices, languages, and experiences. Stories are how we share those. I'm really about stories and memories, which are central in this work of kinship for me. Some memories, I've learned, are genetic. For example, when I eat corn tortillas, I know in my body and spirit that this roots me back to my ancestors and the mountains of Michoacán.

John: With this collective understanding of kinship in mind, how do we become kin? What is the practice of becoming kin? In other words, when we take seriously Robin's point about kin being a verb or process and when we see the world from the perspective of the act of "kinning" across all human and more-than-human scales of the earth and cosmos, how are we to live?

Sharon: The issue of practice to me as a psychologist and mythologist is very much about the imagination. It's about using the imagination and using that human power of mythmaking. For example, there is an old tree on some new land that we have, which is a rowan tree. They tend to grow in cracks and the most unexpected places. If there's a tree that shouldn't be where it is, it's almost always a rowan. It has these beautiful berries and it's beloved of all kinds of birds. And when we moved here in March, that tree looked dead. Its body was more hole than tree.

I said to my husband, "That tree is dead. It's never going to survive." Yet in spring, it suddenly started to develop all of these wonderful shoots coming out of places where trees shouldn't have shoots. Up and down the trunk, as well as at the roots. It was like it was wearing green skirts. Then in autumn, it had an abundance of bright red berries. We made jelly from it.

I've developed a couple of autoimmune diseases just as I'm about to turn sixty. I'm also writing a book on elderhood, and the messages in stories of female elderhood. I was feeling a little bit hopeless during the time when the rowan seemed dead. But then all of sudden there's this old lady tree, as I think of her, that after looking dead in spring comes into her greening, and she's just sitting there laughing at the whole process. She's my inspiration. I call her "The Entwife" because she looks like Tree Beard in *Lord of the Rings*.

To me, it's important that I know a rowan as a tree. I know where it grows. I know what to do with its berries. I know what birds like to be there. I know what other insect life will be there. On a kind of ecological level, I suppose I understand Rowan, but I also see it as something else. That's how I build the relationship with it. It's an act of imagining and a lot of people would call that anthropomorphizing. I don't really believe in anthropomorphizing, because we can only possibly approach a tree as humans. So, you know, what you're looking at is, what is in the space between me and that tree? But, whereas I could quite happily have a conversation with a rowan every day, now I walk past and I have a conversation with the Entwife, and we share stories about what it is to be aging females who have bits of us falling off, and kind of breaking, and holes in us as we get older.

That to me is kind of that imaginal mythmaking ability of humans. It's one of the ways that we can draw ourselves into kinship with other beings that we don't really understand.

Enrique: As María was speaking a little bit ago, and Sharon now, it reminded me of something that we are actively doing right now on our campus at California State University, East Bay. I'm very fortunate to be a professor on one of the most diverse campuses in North America, and also part of an ethnic studies department that celebrates everything that we're talking about. It encourages me as an American Indian studies professor to bring my experiences and those of my students into the classroom.

California just passed a bill, AB 1460, which requires every student in the CSU system to take at least one ethnic studies class before they graduate. The CSU system, and I am on the committee, is tasked with laying out the learning outcomes. Something that we just passed yesterday, that we all decided was important, is (and María reminds me) that the experiences of the students need to be celebrated. Like a couple of you have said, we're often asked to check ourselves at the door, especially our students, when we come into a public space or in a place like a classroom. This new learning outcome encourages students to bring themselves into the classroom. By doing that, we both celebrate and validate the students, creating and validating a larger kinship space in each class and across the classes of twenty-three campuses. This is a living practice of kinship, I believe, a small little seed that I think can grow across the system.

I first need to introduce students to what it *feels* to be in kinship. So many students are born and raised into this modern system with cell phones, Twitter, Instagram, and so on, where they don't really get to experience things in life that connect them with a deeper kinship. In addition to introducing themselves in the classroom with their story, I ask each student to watch either a sunset or a sunrise once a week. They must go to the same spot, the same time of day, once a week, to watch either sunrise or sunset.

Most of them balk at it at first, but then as I read their weekly journals, I get to read their awareness just grow. Some of them do it under the San Mateo bridge, which isn't too far from our campus.

Some of them go on one of the tops of the hills on our campus or in downtown Oakland. Wherever it happens to be, it is amazing to watch their awareness expand, and they're not realizing it, but they are experiencing direct kinship with at least one space in their world. That's my contribution right now, introducing and increasing this practice.

María Isabel: I want to connect something with what you and Sharon said about validating. When you invite the whole self, it's also an act of humanizing all of us, right? We invite everybody to come in and we are vulnerable. When we share our stories, we're creating a more human space and I think that's pretty radical in this world that intends to mechanize us. We engage with the world in so many different ways. All of our stories, not just in terms of our multicultural identities, but all the different ways that each person sees the world.

Sharon, I really appreciate the story of the rowan tree. I haven't seen a rowan tree, and I really want to now look into what it looks like. I think about some of the stories I often share in my classroom that include stories of conversations I have with my nephew and my niece. I don't have any children, but my nephew, my niece, they teach me a lot. One of the coolest things that I love to talk with my nephew about is his belief in the spirits of the trees. He watches all sorts of things, and we have these really cool conversations, and he tells me about our different spirits in the spiritual world. As adults, we forget that this is actual wisdom. Bringing those stories and bringing the diverse ways we engage with each other and each other's worldview into the classroom is another example of how to bring that whole self in.

It's not just saying, "I am a Chicana." I am this, which is my identity, but it's also like, "This is my worldview." Like, I learned from stories. I learned from being in relationship. I will go to the forest, and every single time I will hug a tree because it gives me energy. That's the sort of story that I will bring to class without feeling apologetic about it.

I think a lot of spaces, academic spaces, talking about the spiritual, talking about these other identities, is unaccepted. I am diverse. I speak two languages. But also these are my stories, this is who I am as a spiritual being, this is where I learn and where I take knowledge. I invite students to do the same (to bring their spiritual selves into the classroom). I think it is validating and humanizing, and it also is such a radical thing to do within the insistence that we should compete with each other in an individualist way, right? It's humanizing, it feels good, and it is an active resistance.

Orrin: Thanks for that. One of the reasons we started the podcast—up soon under the name *The Watering Hole*—is related to what you were saying. The roots, the watering hole, I mean, being rooted in our shared cultural history which goes back to the ancestral watering hole and its importance in terms of where we transmitted culture.

I'm surrounded by books; everybody's got books. We spend a lot of time in that space, but this connection with all folks, whatever culture we come from, is rooted in oral tradition and oral transmission. It passes on how to do certain things, right?

In West Africa there's a term, it's a French term called *griot*, for someone who carries an oral tradition. In a podcast, we wanted to relate to that and come at folks by returning to that oral tradition, which is the history of probably 99 percent of our human history. It used to be the written script and being able to read was confined to the priestly classes or the royal classes. And then you had, of course, the printing press and then more people started to read, but it wasn't until a few decades ago that more than half of the people on this planet were "literate." Still only about 80 percent. The way we really relate to each other is by speaking, right?

And it goes beyond that. We can meet each other and not speak the same language and relate energetically from the heart, you know—the love piece. I remember when I was in the air force. I was stationed in Thailand and I knew a little guy I used to call

"Papasan" (I don't even remember his real name now), but he didn't speak a word of English and I spoke a little Thai. He and I spoke a little bit of the time, but I would sit with him every day. You know, we only saw each other for a few minutes and we would not even really speak, but we just liked each other. Right as soon as he saw me, as soon as I saw him, we just felt this *kinship*, if you will. I remember giving him some money as I was leaving Thailand in July 1974. It was not a lot of money. It might've been like fifty bucks, but somebody told me then, "You know, you just fed his family for like six months."

All of these ways for us to communicate is really important to me, and digging back into where everybody comes from, because it really is a universal way of doing things. Connecting back to María's family immigrant story, I had this thing about people of African descent in this country, too: *being in motion*. Being brought to this continent and since then being in motion through things like the Great Migration, which brings folks like me to Chicago. Then you get into these situations with these cities saying things like, "Oh, you got to move because now we're going to gentrify, you know, and you don't fit the profile of what we want here," either economically or racially or whatever the case may be. Right. And it's like, damn, you know, moving again and motion again.

I am saddened because I don't have that place to go to. You have Africa, you have the South in some ways. My mother's ninety, almost ninety-six. She's right down the hall, right? And she's always like, "Oh, I want to go back to Florida," but she was born here in Chicago. But Florida's where her mother and father were born, and we would visit my great-grandmother there. She was just a little African Seminole woman who was like five feet tall. I remember she took this bun down from her hair and it fell down past her waist. We would go down there in the summer and *never* wanted to come back to Chicago. I wanted to stay.

You know, it's about being centered, knowing that wherever I lay my head is home.

John: One part of your essay is very moving—when you talk about going home to Florida and experiencing your first moment of witnessing segregation. In also thinking of your grandmother as Seminole, at one point they faced both the Fugitive Slave Law and the Indian Removal Act, so I wonder if there is a resistance tradition that comes out of being Seminole. It really makes me think about the practice of kinship.

Each of us may be contemplating our version of Enrique's assignment to go watch a sunset. I also suspect (and tell me if I'm wrong) that the practice of kinship involves a level of resistance against those barriers from feeling a sense of safety in terms of where you lay your head.

Orrin: Yeah. Relating to that, my great-grandmother was this little woman who was dark-skinned as well. She's from Pensacola, Florida, that's where they live. But Pensacola, I just found out a year or so ago has this history of being the center for the cohesion and collaboration between African folk that had left plantations and the Native folks from the area. I'm finding that some people of African descent in Mexico actually left from places like Pensacola, in Florida, and went to Mexico, where they became part of the formation of Afro-Mexican communities in Mexico.

Robin: You know, John, your question that gathered us—"How then shall we live?"—has been on my mind in a number of other realms, because I'm right now studying some Anishinaabe teachings that we call the seven grandfather teachings. I've heard them for a long time and understood them as about the virtues that lead to a good life, a good balanced life in relationship.

The teachings were about courage and honesty and love and wisdom and respect. I always heard them as how we should be with one another in order to create the relationships that create a whole life, which is worthy of the gift of breath, honestly. Moreover, as I've been thinking about them, I think about them as the relations, as these practices that cement our kinship with the more-than-human world as well.

It's for me a remembering that as human people we don't have to invent any kind of new kinship for having kinship with salamanders and trees. We already know. We already know how to do these things, you know? It is compassion and generosity and respect and love and humility and all of these things.

I've been very moved by digging into, What does it mean to be humble before trees? What does it mean to show courage to algae? I've been really thinking about this and realizing that for me (at least in the contemporary world, maybe not at the time that the grandfather teachings were first spoken of because it was just a given), in order to engage with these kinship practices with the more-than-human world, the first thing that we need is *attention*. We need to be paying attention in that deep way that we do to each other.

You know, sitting here looking at your beautiful faces and the expressions that pass over your faces as we pay attention and listen deeply to each other—that's where respect comes from, right? That's where kinship comes from. Paying attention, a compassionate kind of attention and empathic attention to the more-than-human world seems a prerequisite for kinship and a practice that we have lost. What Enrique is talking about is regaining the practice—that you're inviting your students to—is to pay attention. When we think about this notion of giving our gifts and receiving the gifts of others as the fundamentals of kinship, I think attention is this wonderful gift that we have as humans. Colonial forces have hijacked our attention to what it is that the more-than-human world may say. We're supposed to be paying attention to productivity, efficiency, status, and ownership, and *that* whole nine yards, but being good kinfolk to me means you're reclaiming our gift of attention and attending compassionately to the other beings. Out of that can come love and respect and humility.

I guess the other thing that's really on my heart, John, is that when we do pay attention—when we come into relationship with each other in a real authentic relationship with each other and with the more-than-human world—we feel such pain. I feel like that's where sometimes, too many times, that leads me to that attentiveness to loss, attention to grief in the world.

"How then, shall we live?" is a question of how, then, if we're paying attention, how do we live with grief? This week being just a microcosm of the grieving we have to do. The question that's on my mind is, How do we act from grief? Because that grief comes from love for the world. How do we move around that circle from attention, that leads to love, the love that leads to grief, and how do we then bring that grief to the practice of kin making?

María Isabel: I love that. I love that idea of attention. I really appreciate that. I love the idea of rekindling our gift of attention, and rekindling and nurturing our right to decide where we want to put our attention. I think there's such a battle around us to claim our attention. Everywhere I look, someone's trying to take my attention and have me look at it.

It's so important to be mindful of that, as you're saying. It is so important to also give our attention to what gives us joy and what brings us energy and what kindles that love that we hold. I think we are, by nature, loving beings. One of the ways, connecting to my nephew and my niece, is that I witness the ways that there's all these things trying to call their little minds, like, "Look at me, look at me." The most joyful moments that I can share with them is sitting with them and being present in that moment and giving them my attention and modeling to *them* what it means to *feel* somebody's attention. When they give me their attention, then there's these moments where they're looking around and I'm like, "Hey, nephew, can I see your eyes?" And he looks at me, he smiles. He says, "Yes, Tía, I'm here." Right. Just doing that and doing small moments of giving each other our loving attention and being present is so important, because the corporate world is trying so hard to rob our young kids of their attention, trying to sell everywhere—YouTube, everything six-year-olds don't know how to do—all these things in ads *everywhere*. It is so overwhelming. It is so important to note what we give our attention to. Who do we give our attention to and how do we model what attention, compassionate attention and being present, means?

Robin, I want to just say thank you for that. I also want to add something to the question of what it means to live a good life: cling to a right to experience joy. I think joy is different from happiness. We're not happy all the time. You know, certainly I'm not, I'm pretty enraged a lot of times. But I think joy is something else. Joy is part of being whole, in here in my heart I am feeling OK and complete. We feel joy when we cling to our dignities. Sometimes our dignities can feel enraged. But it's righteous and it's OK. But we can't lose that joyful nature that we have. To me, it's clinging as humans, as beings, as folks in community, to joy. Clinging to all our relations that are not human. I wanted to add joy to that gift of attention.

John: That's attention, right? Attention expands kinship across the human community and into the more-than-human realm that sustains us. In terms of the practice of attention I think about my daughters. I'll often say to them something which is easier said than done. I'll say, "There's listening to think about what you're going to say next. And then there's listening for the sake of what's being said," and the second is harder.

Silence in general, especially attentive silence, is harder. I'm reminded of a time when I was deeply humbled on the topic of silence. It was my first time trekking in the Himalayas, in India above Gangotri Glacier, where the first trickles of the Ganges begin. I saw a hole in a rock and realized it was a cave. A man stepped out and invited me up. He was a Swami taking a vow of silence, but he could write in English and he'd been there for years. I spent two days with him. Me talking; him writing. He would ask me a question and when I would answer he would write, "Quiet is better." Over and over. No matter what I said. He wanted to have a conversation. When he found out I was a Western philosopher, he was really excited to talk, kept cooking me rice and tea and hoping I'd stay. But whenever I'd really say something that in my Western ego I thought was profound, he would write, "Quiet is better." I returned to that town of Gangotri below the glacier and I bought this bracelet [*holding it up to the Zoom camera*] and whenever I really need to pay attention (which is hard with my American, male, Irish, Catholic, Jersey Boy background, and as someone who just wants to jump in and celebrate people's ideas by talking over them—my form of attention from my upbringing), I'll just spin the beads on this bracelet and think, "Quiet is better."

I try and try to make sure I'm not just listening for what I'm going to say next. But that next level of attention is truly humbling—what I think Sharon's talking about with really understanding a tree in itself, not what it does to serve us—or Orrin's practice of interviewing in his podcast, or Enrique's observation of the sunset, or María Isabel's clinging to joy when totally focused on her niece and nephew.

So, in closing, is there anything we want the reader to just attempt as a "practice of kinship"? Y'all are such humble people and I respect that you don't want to say "here's how you should live" to our audience, but is there a practice that people can just try on for size each day and see if it fits?

Orrin: For me, it is gardening. Like with what Robin said, when you garden, you pay attention, right? It draws you into paying attention because, you know, what are you going to plant now? You look at what you're growing and you're trying to see if the soil or seeds or plants are stressed in any way, if something's bothering them. You know, when I get up in the morning and go out, I say, "I'm going to pay attention to my children, these plants." We want to expand this practice in terms of backyards here in Chicago, but that's one thing I would recommend people to do.

And don't give up! It's like, that's the other thing: we're here to help. The seasons are longer than you think. Once they see that, they pay attention to the earlier appearance of the spring and may work late into the fall. You know, and now you're paying attention to the sunlight. You've been paying attention to where that sunlight is in your yard. Now you're paying attention to whether your yard is sunny or whatever the case may be. And now you notice that the pollinators have to have flowers, and you notice that there's different kinds of wasps and different kinds of bees and all of these sorts of things. So that's one thing I would recommend for folks: garden. Particularly growing something you can eat, which changes our houses, our buildings, our residences from centers of consumption to places where we're also producing.

John: It also brings our tongue, our tastes, our stomach, our nutrients, our health into all the layers of kinship all of you just talked about, from soils to plants to people to sun to cosmos. Thank you for that one, Orrin.

Enrique: In terms of daily practice, I have two things. One of the things I taught to you, John, and your folks over there at the Headwaters Conference at Western Colorado University: my "Story of Self, Story of Us, Story of Now" workshop turns story into the practice of kinship. One of the classes I teach is American Indian Oral Literature. It's not just about story*telling*. It's about story*listening*, as you saw in the workshop at Western.

One of the things I have my students do, and I suggest this to everyone: learn how to engage in active listening. I have my students pair off—and you can do this with a rock, do this with a tree, or your dog—and for two minutes, just listen without any "I," meaning you remove yourself. You just listen.

In my classrooms, I have my students do this for ten minutes and then they switch off and then afterward I ask them to share what the other person told them. Active listening is a skill. It's tough these days because as María pointed out, you know, we're drawn in ten different directions. It's tough to actually just actively listen to one thing these days. That could be a very powerful practice.

The other thing is related to what I mentioned earlier. You reminded me of one of my students in that sunset practice. I was reading his journals afterward, and the first three weeks or so he was wondering, you know, "Why are we even doing this?" Then at one point in his practice, he noticed a deer show up to the area where he was watching the sunset. Then he noticed the deer eating one of the bushes. And it happened every week. He got curious, and first decided to do research. What kind of deer was that? And he started to research. What was the bush? Then he even started to research more—what part of the bush was the deer chewing on? Throughout the semester, through his increasing curiosity, he developed a kinship with that one little space where he had done his sunset practice.

I guess my suggestion is to develop this curiosity. We're not as curious about things around us these days. We have to develop a curiosity, to expand our awareness, to expand kinship with everything around us. Like Orrin was saying about gardening—say, you grow some jalapeños in your garden. You gotta make sure that the jalapeños have sun, gotta make sure the soil isn't too wet. Then you start to pay attention to other things you want to add to your jalapeños. Maybe you wanna grow some cilantro or some other things. The curiosity keeps growing. It's just amazing how our awareness and our kinship expands by one little simple practice.

John: I think it brings us back to Robin's point that when it comes to the practice of kinship, always and already, we are really just recollecting, reawakening, or just noticing what was already there when we return to attention, gardening, curiosity, kinship. I love the idea of curiosity as a portal into the kinship that's already there. I think that is beautiful. We are not "discovering" anything.

Sharon: Every time I go outside, I speak to things I encounter. I address them. Even if it's just a simple "Hello, crow." I talk to trees,

I talk to rocks, to flowers. I listen for anything they have to say back to me, of course, but mostly I pay attention to them. If I'm sitting by the spring well on our land, I'll sing to it. Maybe offer a line from a poem that particularly grabbed me that morning to the Entwife. That way, I'm not seeing the world as Other. I'm seeing it as something I'm in relationship with, seeing all the beings around me as beings I'm in relationship with. What you speak to, you wake up. Part of the practice of kinship is maybe to help keep the world—or the ancient tradition of the world soul—awake.

María Isabel: Play. I just want to add play as a practice of kinship. Play! Just go out and play, whatever that looks like. There is so much serious work all the time, you know? But I think it's important to try to rekindle that we are by nature playful beings. Whether you play an instrument—I see all those instruments behind Enrique—or I run around chasing my dog. My dog loves his favorite game of running around the house, and we run around in circles. Or as Orrin and Enrique said, getting our hands dirty from playing in the garden.

Orrin: [*Holding up a wooden flute*] Yes! This is my COVID thing. Oh, my God. Loving music, and this flute is Native American, but let's learn to play with music and flutes from different cultures, right? Let's try that. So play, absolutely María.

John: Kinship through play. What a great last word. Thank you all. What wonderful practices of kinship: *attention, gardening, curiosity, play*.

Robin: Can I add a last word to the last word? Out of everything you are all sharing that come from play and attentive care, I think a practice that can help really cement the kinship is to flow from that attention to gratitude. Let's translate when we see and when we pay attention into the big, radical kind of gratitude for all of those gifts.

[*All nodding, in radical gratitude, to all.*]

PERMISSIONS

These credits are listed in the order in which the relevant contributions appear in the book.

"After" was originally published in Heather Swan, *A Kinship with Ash* (West Caldwell, NJ: Terrapin Books, 2020).

"Mercy" is used with the permission of Nickole Brown. © 2018 by Nickole Brown, *To Those Who Were Our First Gods* (Rattle, 2018).

ACKNOWLEDGMENTS

As editors, we want to offer some deeply felt gratitude. An initial gathering in 2018—enabled by the generosity of the Center for Humans and Nature—began the conversations that would eventually become the *Kinship* series. During our time together, our group of twenty or so people often sat in a loose circle—some in chairs, some leaning against couches, some cross-legged on the floor—listening to unique experiences from all varieties of places. About this gathering, mostly it is the joy we remember, the laughter that seemed to bubble up spontaneously, reinforcing bonds of kinship and love for this living earth as we attempted to put this into words and actions.

We also recall the ways other voices came into our midst. During one of the meetings, for example, a participant in the middle of the circle pulled out her iPhone and gently asked us to listen. The sound of a family of Orcas communicating with one another entered the room. Even those of us who don't live anywhere near the Pacific Ocean felt deep recognition, made more poignant by the current threats faced by these fellow mammals. They were speaking to one another, yet it felt as though their voices were also reaching through the water to us. Kin.

Later, we ventured outside to stretch our legs and to participate in a soundwalk to attune ourselves to the nearby forest and its aural textures. We paused at a huge Oak whose sprawling crown seemed to cover most of the backyard. In a way, this long-lived Oak was the reason we gathered where we had. This tree's presence preceded the home in which we were meeting by many decades, likely centuries, and the home was there because the parents of

Strachan Donnelley, the founder of the Center for Humans and Nature, chose to live in this spot. The family story has it that this elder Oak tree drew them to the place. We paid our respects. Some of us laid a hand on the tree. All of us breathed, thinking about plant kin who outlive us in age and who gift us with oxygen.

On the basis of this meeting, the initial vision for *Kinship: Belonging in a World of Relations* was the creation of a single volume. Then it grew. As each of us, as editors, reached out to people of various expertise, asking them to share their perspectives and stories of kinship, new threads were suggested to us and the web became larger. Soon it was apparent that the web could not be contained by a single volume—at least one that abided by the rules of standard publishing specs. We decided to ask, What if? What if we let form follow function? What if we let this book become what it wants to become? What if this book should be a series? This line of questioning begat logistical problems. For one, publishers in an already-risky landscape can't comfortably take those kinds of chances.

Yet challenges can also create innovations. The *Kinship* series has thus become the Center for Humans and Nature's first venture into book publishing—of many, we hope. We're thrilled with the outcome. We'd like to express our deep gratitude for human kin that helped bring this ambitious project to such a beautiful result. Emily Lonigro, Demetrio Cardona-Maguigad, Felix Castellanos, Carla Levy, and Stacey Saunders of LimeRed, for the gorgeous book covers and visual design. Minds blown. Manuscript editor Katherine Faydash, for her expert eye and enthusiastic support of the content. Riley Brady, for the lovely page layout and design. Ronald Mocerino at the Graphic Arts Studio Inc., for meeting our every printing need. Chelsea Green Publishers, especially Michael Weaver and Michael Metivier, for being our fine confidants in distribution and promotion. Paul and Sandy Quinn, for hosting us so graciously at Windblown Hill. For their giant spirits, supportive presence, and care-filled work, the Center for Humans and Nature kinfolk: James Ballowe, Hannah Burnett, Anja Claus, Katherine Kassouf Cummings, Jon Daniels, Brooke Hecht, Bruce Jennings, Curt Meine, and Jeremy Ohmes; and for the support of the Center for Humans and Nature board: Gerald Adelmann, Julia Antonatos, Jake Berlin, Ceara Donnelley, Tagen Donnelley, Kim Elliman, Christopher Getman, Charles Lane, Thomas Lovejoy, Ed Miller, George Ranney, Bryan Rowley, and Eleanor Sterling.

Gavin would like to give extra thanks and love to his family—Marcy, Hawkins, and Peanut—for indulging him as a "nature nerd" and for keeping his heart full. Also, Coyote, Magpie, and the Crab-like Orbweaver—the world wouldn't be the same without you.

Robin offers gratitude to her human and more-than-human kinfolk for their loving support: Family, Students, Maples, Orioles, Foxes, Peepers, and the whole dazzling web of relations.

John would like to thank his wife, Karen, for her near quarter century of partnership in cultivating kinship, from raising our daughters Atalaya and Sol to sharing work as teachers in the School of Environment and Sustainability at Western Colorado University to building Camp Alpenglow in the heart of the Gunnison Country. None of John's work is possible without the mountains and snows that frame and stand sentinel above his valley, providing greater-than-human family since he first visited from New Jersey at age sixteen.

CONTRIBUTORS & KIN, VOLUME 5

Dr. Sharon Blackie is an award-winning writer and internationally recognized teacher whose work sits at the interface of psychology, mythology, and ecology. Her highly acclaimed books, courses, lectures, and workshops are focused on the development of the mythic imagination and on the relevance of our native myths, fairy tales, and folk traditions to the personal, social, and environmental problems we face today. As well as writing four books of fiction and nonfiction, including the best-selling *If Women Rose Rooted*, her writing has appeared in the *Guardian*, the *Irish Times*, the *Scotsman*, and more, and she has been interviewed by the BBC and other major broadcasters on her areas of expertise.

Nickole Brown is the author of *Sister* and *Fanny Says*. She lives in Asheville, North Carolina, where she periodically volunteers at several different animal sanctuaries. Her work speaks in a queer, Southern-trash-talking way about nature beautiful, damaged, and in desperate need of saving. The first collection of these poems, *To Those Who Were Our First Gods*, won the 2018 Rattle Chapbook Prize, and in 2020, her essay-in-poems, *The Donkey Elegies*, was published. In 2021, Spruce Books of Penguin Random House published *Write It! 100 Poetry Prompts to Inspire*, a book she coauthored with her wife, Jessica Jacobs.

Sunil Chauhan is a social innovator and a nature wellness expert. His journey began in the high Himalayas amid pristine forests of oak, cedar, spruce, chestnut, and other temperate trees, flora, and fauna beside the mighty Himalayan peaks and rivers. He considers the woods to be his school, university, ashram, and home. His formal education includes degrees in philosophy and history as well as in information technology. He has studied Eastern mysticism and is a believer in the tradition of biodivinity.

Alison Hawthorne Deming's most recent books are the poetry collection *Stairway to Heaven* (2016) and the essay collection *Zoologies: On Animals and the Human Spirit* (2014). Her Guggenheim Fellowship book *A Woven World: On Fashion, Fishermen, and the Sardine Dress* will be published by Counterpoint in 2021. She is Regents Professor at the University of Arizona.

Photo by Bear Guerra

Thomas Lowe Fleischner is Senior Advisor & Director Emeritus of the Natural History Institute (naturalhistoryinstitute.org) in Prescott, Arizona, Faculty Emeritus at Prescott College, Past Chair of the Natural History Section of the Ecological Society of America, and a Fellow of the Linnean Society of London. He is the editor of two anthologies: *The Way of Natural History* and *Nature, Love, Medicine: Essays on Wildness and Wellness*, and author of two other books and numerous articles. (tfleischner.net)

Tiokasin Ghosthorse—Cheyenne River Lakota Nation of South Dakota—is an international speaker on peace, Indigenous, and Mother Earth perspectives. A survivor of the "Reign of Terror" from 1972 to 1976 on the Pine Ridge, Cheyenne River, and Rosebud Lakota Reservations in South Dakota and the US Bureau of Indian Affairs Boarding and Church Missionary School systems designed to "kill the Indian and save the man," Tiokasin has a long history of Indigenous activism and advocacy. He spoke as a fifteen-year-old at the United Nations, in Geneva, Switzerland. Tiokasin was a 2016 nominee for the Nobel Peace Prize from the International Institute of Peace Studies and Global Philosophy.

Photo by Ivan March

Matthew Hall originally trained as a botanist at the University of Edinburgh, and he worked for several years as a research scientist at the Royal Botanic Garden Edinburgh. There he cofounded a research center and worked on dozens of research and conservation projects across the Middle East, including in Afghanistan, Iraq, Libya, and Yemen. After gaining his PhD in environmental philosophy from the Australian National University, he published *Plants as Persons: A Philosophical Botany* (SUNY Press, 2011). Matthew's new book, *The Imagination of Plants: A Book of Botanical Mythology* (SUNY Press, 2019), examines how the world's mythological traditions perceive and relate to the plant kingdom.

John Hausdoerffer, jhausdoerffer.com, is author of *Catlin's Lament: Indians, Manifest Destiny, and the Ethics of Nature*, as well as coauthor and coeditor of *Wildness: Relations of People and Place* and *What Kind of Ancestor Do You Want to Be?* John is dean of the School of Environment & Sustainability at Western Colorado University and cofounder of Coldharbour Institute, the Center for Mountain Transitions, and the Resilience Studies Consortium. John serves as a fellow and senior scholar for the Center for Humans and Nature.

Photo by Keith Carlsen Photography

Trebbe Johnson is the author of *Radical Joy for Hard Times: Finding Meaning and Making Beauty in Earth's Broken Places*, *The World Is a Waiting Lover*, and *101 Ways to Make Guerrilla Beauty*. She is also the founder and director of the global community Radical Joy for Hard Times, devoted to finding and making beauty in wounded places. Trebbe speaks four languages; has camped alone in the Arctic wilderness; studied classical Indian dance; and worked as an artist's model, a street sweeper in an English village, and an award-winning multimedia producer. Her forthcoming book is *Fierce Consciousness: Surviving the Sorrows of Earth and Self*. She lives in Ithaca, New York.

Dr. Robin Kimmerer is a mother, botanist, writer, and Distinguished Teaching Professor at the SUNY College of Environmental Science and Forestry in Syracuse, New York, and the founding director of the Center for Native Peoples and the Environment. She is an enrolled member of the Citizen Potawatomi Nation and a student of the plant nations. Her writings include *Gathering Moss* and *Braiding*

Sweetgrass: Indigenous Wisdom, Scientific Knowledge and the Teachings of Plants. As a writer and a scientist, her interests include not only restoration of ecological communities but also restoration of our relationships to land. She lives on an old farm in upstate New York, tending gardens domestic and wild.

María Isabel Morales, professor at the Evergreen State College, received her PhD in cultural studies and social thought in education from Washington State University. Born in Michoacán, Mexico, Dr. Morales draws from her family's stories of (im)migration and cultural knowledge in her teaching, research, and community engagement. As a bilingual queer Mexicana woman, Dr. Morales knows what it feels like to "live in the borderlands" and uses this conocimiento to contribute to a pedagogy of love and human dignity. Morales teaches in cultural studies in education, Latinx studies, and political economy.

Ajay Rastogi is cofounder of the Foundation for Contemplation of Nature and director of the Vrikshalaya Himalayan Centre in India. He is the recipient of the South Asian Youth Leader award, an Erasmus Mundus Fellow, and a Nehru Fulbright Environmental Leadership award. He was involved in designing a protected areas network, a Himalayan ethnobotany initiative in five countries, and an FAO (UN) Organic Agriculture Programme in India. Now, he facilitates Resilient Leadership in the Himalayas, a course in association with a Jagriti women's self-help group. They won an International Mountain Award (2020) for transformative education. His e-book *Contemplación de la naturaleza* is available in Spanish.

Photo by Lauren Dragons

Photo by Brooke Hummer

Jill Riddell is host of *The Shape of the World*, a podcast about nature and cities. She is founder and operator of the Office of Modern Composition and serves on faculty at the School of the Art Institute of Chicago, where she teaches in the creative writing program.

Enrique Salmón is a Rarámuri (Tarahumara) Indian. He has a PhD in anthropology from Arizona State University. He is head of the American Indian Studies program at Cal State University East Bay. He has been a scholar in residence at the Heard Museum and has served as a board member for the Society of Ethnobiology. He has published many articles on indigenous ethnobotany, agriculture, nutrition, and traditional ecological knowledge. Dr. Salmón is author of the books *Eating the Landscape: American Indian Stories of Food, Identity, and Resilience* and *Iwígara: The Kinship of Plants and People*.

Amba J. Sepie, PhD, is a writer, researcher, and teacher who works in the field of human and cultural geography at Te Whare Wānanga o Waitaha, Aotearoa New Zealand. Her award-winning doctoral thesis, "Tracing the Motherline," profiles the offerings of global Indigenous Elders, scholars, and communities in conversation with strategies for decolonization and socio-ecological healing, themes that are extended in her various publications.

Heather Swan is the author of the poetry collection *A Kinship with Ash* (Terrapin Books) and the nonfiction book *Where Honeybees Thrive: Stories from the Field* (Penn State Press), winner of the Sigurd F. Olson Nature Writing Award. Her nonfiction has appeared in many publications, including *Aeon*, *Belt*, *Catapult*, and *Minding Nature*. Her poetry has appeared in *Terrain*, *Poet Lore*, *The Hopper*, *Phoebe*, *Cold Mountain Review*, *Midwestern Gothic*, *Wildness*, and several anthologies. She has been the recipient of the Martha Meyer Renk Fellowship in Poetry, the August Derleth Prize, and an Illinois Art Council Fellowship. She teaches writing and environmental literature at University of Wisconsin–Madison.

Gavin Van Horn is the creative director and executive editor for the Center for Humans and Nature. His writing is tangled up in the ongoing conversation between humans, our nonhuman kin, and the animate landscape. He is the co-editor (with John Hausdoerffer) of *Wildness: Relations of People and Place*, and (with Dave Aftandilian) *City Creatures: Animal Encounters in the Chicago Wilderness*, and the author of *The Way of Coyote: Shared Journeys in the Urban Wilds*.

Maya Ward is passionate about deepening the connections between body, ecology, and culture through writing, dancing, and tending the earth. She has a PhD in creative writing. Her book *The Comfort of Water* is an account of her twenty-one-day journey from the sea to the source of the Yarra River, following the length of a Wurundjeri Songline. Currently she lives on the banks of the Yarra and cocreates pilgrimage-based ritual, teaches dance, shares her knowledge through public speaking, and grows food with her home community in the mountain village of Warburton. www.mayaward.com.au

Kyle Whyte is George Willis Pack Professor of Environment and Sustainability at the University of Michigan, serving as a faculty member of the environmental justice specialization. Previously, Kyle was professor and Timnick Chair in the Department of Philosophy and Department of Community Sustainability at Michigan State University. Kyle's research addresses moral and political issues concerning climate policy and Indigenous peoples, the ethics of cooperative relationships between Indigenous peoples and science organizations, and problems of Indigenous justice in public and academic discussions of food sovereignty, environmental justice, and the Anthropocene. He is an enrolled member of the Citizen Potawatomi Nation.

Orrin Williams's work-life is as the food systems coordinator for the Chicago Partnership for Health Promotion, a program of the Office of Community Engagement and Neighborhood Health Partnership at University of Illinois at Chicago. Otherwise, he is a student of spiritual matters, particularly those related to acknowledging our relationships with Oneness and all beings. Orrin does that through gardening, reading, research, writing, and as the executive director of the Center for Urban Transformation in Chicago and the *Roots Watering Hole* podcast series with cohost and soil scientist Dr. Akilah Martin.

 **Anthony Zaragoza** is the son, grandson, and great-grandson of steelworkers and pipefitters. He has worked at the Evergreen State College since 2004. Zaragoza has been teaching and learning political economy, popular education, and cultural studies inside and outside academia, including inside women's, men's, and youth prisons as well as various movement spaces. He has received various grants for his project Neoliberalism in the Neighborhood, which examines economic, political, and social changes in communities over the past fifty years, including his own neighborhoods in Hammond, East Chicago, and Gary, Indiana.